A partial synopsis of the *Euphorbiaceae-Platylobeae* of Australia (excluding *Phyllanthus*, *Euphorbia* and *Calycopeplus*)

H. K. Airy Shaw

Summary. A partial synopsis of the *Euphorbiaceae-Platylobeae* of Australia is offered. The *Stenolobeae*, endemic to Australia, are omitted as are accounts of the species of *Calycopeplus*, *Euphorbia* and *Phyllanthus*. *Synostemon* is sunk into *Sauropus* and the 18 new recombinations made; other new combinations are *Adriana glabrata* var. *subglabra*, *Croton stockeri* and *Drypetes lasiogyna* var. *australasica*; *Sauropus albiflorus* subsp. *microcladus* is a stat. nov.; new taxa: *Sauropus hubbardii* and *S. latzii*, *Glochidion sessiliflorum* vars. *pedicellatum* & *stylosum*, *Sauropus brunonis* var. *ovatus*, *S. elachophyllus* var. *glaber* and var. *latior S. glaucus* var. *glaber*.

When I embarked upon this enumeration a few years ago, I had hoped that the excluded genera mentioned above would be tackled by other workers who I knew were interested in them. This hope has not yet been realized, and since advancing years and the difficulties of travel make it impracticable for me to commence the study of these groups with any prospect of adequate coverage or ultimate completion, I have decided, with regret, to publish the remainder in its present incomplete state.

The enumeration represents a logical extension of the accounts already published for Siam, Borneo and New Guinea; it is indeed evident that the Australian Euphorbiaceous flora (the endemic *Stenolobeae* excepted) is largely a southerly extension of that of Malesia. The relative numerical representation is, however, mostly very different. For example, the two most numerous genera in New Guinea, *Macaranga* and *Glochidion*, comprising there some 70 species each, are represented in Australia by a mere 7 and 15 respectively. Other similar examples are provided by *Claoxylon*: New Guinea 26, Australia 3; and *Mallotus*: New Guinea 20, Australia 12. *Phyllanthus* (*s. str.*, excluding *Synostemon*), the most numerous genus in Australia, with about 45 species, has some 45–50 in New Guinea. The respective representations of *Croton* in the two areas are also approximately equal, with about 20 species apiece.

Australia has a few small endemic or near-endemic genera: *Adriana* (4–5 spp.), *Petalostigma* (6 spp., 1 just spilling over into Papua), *Dissiliaria* (3), *Calycopeplus* (3), *Neoroepera* (2), *Rockinghamia* (2), *Hylandia* (1), *Whyanbeelia* (1). On the other hand, a number of genera, widespread in Malesia and represented in New Guinea, appear not to have reached Australia, or if they once did so are now extinct there: for example, *Aporusa*, *Baccaurea*, *Blumeodendron*, *Agrostistachys*, *Ptychopyxis*, *Melanolepis*, *Neoscortechinia*, *Spathiostemon*, and others.

Accepted for publication February 1980.

This paper is obtainable separately from the Royal Botanic Gardens, Kew, Richmond, Surrey, England; price £6, postage and packing extra. It is the latest of a series by Mr Airy Shaw which began with the Euphorbiaceae of Siam (Kew Bulletin, Vol. 26, No. 2 (1971)), also available from the Royal Botanic Gardens, price £5, the Euphorbiaceae of Borneo, Kew Bulletin Additional Series IV, available from HMSO price £10, and the Euphorbiaceae of New Guinea, available from HMSO price £15.

MAP 1. Map of Australia showing subdivisions as used in this paper. Names of states omitted.

Representatives of some of these may yet turn up in unexplored parts of Arnhem Land and the Cape York Peninsula. The genera *Margaritaria*, *Cleidion*, *Trigonostemon*, *Suregada* and *Pimelodendron* have only been added to the Australian flora within the last 10 years.

It is probably unnecessary to repeat here the general explanatory matter prefaced to my Bornean enumeration (Kew Bull. Add. Ser. IV: 1–2 (1975)), beyond pointing out that the arrangement of both genera and species is alphabetical, but that for those interested in the classification of the family a tentative natural arrangement of genera may be found in the same place, pp. 4–5.

In order to save space, the most frequently cited literature references have been reduced to author (often abbreviated), page and date. The titles and volume numbers should be supplied in each case as follows:

Anderson—Trees of New South Wales, ed. 4 (1968).
Backer & Bakh. f.—Flora of Java 1 (1963).
Bailey—The Queensland Flora 5 (1902).

Beadle, Evans & Carolin—Flora of the Sydney Region, rev. ed. (1972).
Benth(am)—Flora Australiensis 6 (1873).
DC.—De Candolle, Prodromus Syst. Nat. Regni Veget. 15(2) (1866).
Domin—Beiträge zur Flora Australiens, in Bibliotheca Botanica 22 (Heft 89 iv) (1927).
Engl(er)—Pflanzenreich IV. 147 (1910–24).
Ewart & Davies—Flora of the Northern Territory (1917).
Francis (ed. 3, Chippendale)—Australian Rain-Forest Trees (1970).
Gardner—Enumeratio Plantarum Australiae Occidentalis (1931).
Hyland—Card Key Rain Forest Trees North Queensland (1971).
Jabl(onszky)—*Cleistanthus* & *Bridelia* in Engl., Pflanzenr. IV. 147. viii (1915).
K.B.—Kew Bulletin.
K.B. Add. Ser.—Kew Bulletin, Additional Series.
Moore & Betche—Handbook of the Flora of New South Wales (1893).
Muell(er)Arg(oviensis)—*Euphorbiaceae* in DC., Prodr. 15(2) (1866).
Pax *or* Pax & Hoffm(ann)—*Euphorbiaceae* in Engl., Pflanzenr. IV. 147. i–vii, ix–xvii (1910–24).
J.J.Sm(ith)—*Euphorbiaceae* in Koord. & Valet., Bijdr. No. 12 Boomsoorten Java, in Meded. Dep. Landb. No. 10 (1910).
Specht—in Specht & Mountfort (ed.), Rec. Amer.-Austr. Sci. Exped. Arnhem Land 3 (Bot. & Pl. Ecol.) (1958).

In the brief summaries of distribution the abbreviations used for the six mainland states (NSW, NT, Q, SA, V, WA) should be self-explanatory. (Map 1.)

The following abbreviations have been adopted for the pastoral districts of Queensland:

BK—Burke
BT—Burnett
CK—Cook
DD—Darling Downs
GN—Gregory North
GS—Gregory South
LT—Leichhardt
MA—Maranoa
MI—Mitchell
MO—Moreton
NK—North Kennedy
PC—Port Curtis
SK—South Kennedy
WB—Wide Bay
WG—Warrego

The Districts of the Northern Territory are those used by Chippendale, Check List of Northern Territory Plants, in Proc. Linn. Soc. N.S.W. 96(4): 209, map (1972).

The Divisions of New South Wales are those employed by Anderson, Trees of New South Wales, ed. 4, map facing p. 7 (1968).

Cited specimens are at K unless otherwise stated.

It is a pleasure to acknowledge the kind help of Mr Trevor Stanley, of the Queensland Herbarium, in supplying distributional and other details for several species in Queensland and New South Wales.

Thanks are also due to Ann Davies who drew the figures (and the *Hylandia* on the cover of the separate offprint).

KEY TO GENERA OF AUSTRALIAN EUPHORBIACEAE (*Platylobeae* only)

1. Leaves opposite or apparently so:
 2. Stellate or fascicled hairs present (sometimes very small):
  3. Styles branched; anther-connective acute, cellular-papillose **Adriana**
  3. Styles unbranched; connective not acute . . . . . **Mallotus**
 2. No stellate hairs present:
  4. Inflorescence composed of cyathia; milky juice present:
   5. Ovary surrounded by a 4–6-lobed perianth or involucel within the cyathium . . . . **Calycopeplus**
   5. Ovary without a perianth or involucel within the cyathium **Euphorbia**
  4. Inflorescence not composed of cyathia:
   6. Petals present, conspicuous; 1 ovule in each ovary-loculus **Baloghia**
   6. Petals absent; 2 ovules in each ovary-loculus:
    7. Capsule conspicuously 3-horned, with widely separated recurved styles (but tips often broken off); leaves often finely and shallowly crenulate . . . . **Choriceras**
    7. Capsule not 3-horned; styles shortly connate, recurved **Dissiliaria**
    7. Capsule not 3-horned; stigmas shortly cordate-ovate, subsessile **Austrobuxus**
1. Leaves alternate:
 8. Leaves densely or sparsely glandular-granular below (occasionally above also):
  9. Anthers 3–4-locellate . . . . **Macaranga**
  9. Anthers regularly 2-locular . . . . **Mallotus**
 8. Leaves not glandular-granular:
  10. Leaves pellucid-punctate; inflorescence leaf-opposed, congested-cymose or fascicled, ± gummy when young; twigs remaining green when dry . . . . **Suregada**
  10. Leaves not pellucid-punctate; inflorescence not leaf-opposed:
   11. Stellate hairs or scales present (sometimes sparse or minute):
    12. Inflorescence about as broad as long (± diffusely cymose or thyrsoid):
     13. Fruit less than 1 cm diam.; petals absent . . **Endospermum**
     13. Fruit over 1 cm diam.; petals present, conspicuous . **Aleurites**
    12. Inflorescence much longer than broad (spicate, racemose, or very narrowly thyrsoid):
     14. Inflorescence bisexual (occasionally only 1 basal ♀ flower); petals present in ♂ flower . . . . **Croton**
     14. Inflorescence unisexual; petals absent . . . . . **Adriana**
   11. No stellate hairs or scales present:
    15. Leaf-base distinctly asymmetrical (flowers mostly fascicled):
     16. Fruit a dry or fleshy 1–2-locular drupe; tepals of ♂ flower commonly 4, much imbricate . . . . **Drypetes**
     16. Fruit a 3–25-locular capsule or small 3–8-locular berry or drupe; tepals of ♂ flower 5–6:
      17. ♂ flower without a disk; styles ± united, erect or depressed, often pulvinate or cylindric or much reduced, rarely free; ovary up to 25-locular . . . . **Glochidion**
      17. ♂ flower with a disk or glands; styles free or connate below, often bifid, usually spreading; ovary up to 8-locular . **Phyllanthus**

15. Leaf-base symmetrical:
18. Petals present in ♂ flower (sometimes very small):
19. Inflorescence racemose, mostly longer than broad (see 19 & 19 below):
20. Anther-loculi parallel, distinct, extrorse; filaments stout; no suborbicular bract-leaves . . . . **Baloghia**
20. Anther-loculi back to back, confluent at apex; filaments slender; inflorescence arising in the axil of a small suborbicular leaf or bract; foliage leaves very variable in size and shape **Codiaeum**
19. Inflorescence fasciculate, or flowers solitary, axillary (see also 19 below):
21. Calyx-segments clearly valvate; petals small, obovate or spathulate; ovules 2 per loculus:
22. Fruit a 3-locular capsule; lateral nerves rarely distinctly parallel . . . . **Cleistanthus**
22. Fruit a 1–3-locular drupe; lateral nerves often conspicuously parallel . . . . **Bridelia**
21. Calyx-segments imbricate:
23. Petals equalling or exceeding the calyx; ovules 1 per loculus **Trigonostemon**
23. Petals shorter than calyx; ovules 2 per loculus:
24. Trees or shrubs with larger chartaceous or coriaceous leaves; disk-glands episepalous . . . . **Actephila**
24. Subherbaceous undershrubs, with small membranaceous leaves; disk-glands epipetalous . . . . **Leptopus**
19. Inflorescence about as broad as long (± diffusely cymose or thyrsoid):
25. Petals densely white-tomentose all over; calyx cupular, shortly lobed or dentate; fruit a hard, ellipsoid, 3–6-angled drupe **Fontainea**
25. Petals glabrous or variously pubescent but not white-tomentose:
26. Ovary and fruit bilocular; ♂ petals reddish-pilose within; tall tree . . . . **Hylandia**
26. Ovary and fruit trilocular; ♂ petals glabrous; shrubs or treelets:
27. Calyx shortly golden-sericeous; petals white; capsule muricate **Dimorphocalyx**
27. Calyx densely whitish-pubescent; petals pink or deep crimson; capsule smooth . . . . **Trigonostemon**
18. No petals in ♂ flower:
28. Stamens united in phalanges; flowers in terminal bisexual racemes; leaves palmately lobed . . . . . . RICINUS
28. Stamens free, or united in a column:
29. Slender twining shrub, with ovate-triangular leaves, stinging hairs and short androgynous leaf-opposed inflorescences **Tragia**
29. Erect shrubs or trees, without stinging hairs:
30. Fruit an echinate or muriculate capsule:
31. Inflorescences unisexual; leaves opposite or alternate, but not crowded into false whorls; stamens not interspersed with small processes . . . . **Mallotus**

31. Inflorescences bisexual; leaves crowded into false whorls; stamens interspersed with small processes . **Rockinghamia**
30. Fruit an unarmed capsule or drupe:
32. ♂ flower compressed, with 2 suborbicular tepals:
33. Leaves shallowly crenate, coriaceous; stipules minute or obsolete; inflorescences axillary; no glands beneath each flower; fruit large, massive, drupaceous . **Pimelodendron**
33. Leaves entire, membranaceous to chartaceous; stipules usually conspicuous, elongate, membranous; inflorescences terminal; 1–2 glands beneath each flower; fruit small, glaucous, dehiscent or not . . . . **Homalanthus**
32. ♂ flower otherwise:
34. ♂ flowers fascicled, or solitary, axillary (see 34 & 34 below):
35. Stamens 8 or more, free:
36. Stigmas subsessile, ovate, subcordate . . **Austrobuxus**
36. Stigmas ± flabellate:
37. Fruit ultimately dehiscent; pericarp fleshy, wrinkled, shining; stigmas 3–4, elongate, conspicuous, caducous **Petalostigma**
37. Fruit indehiscent; pericarp dry, dull; stigmas 1–2, small, ± persistent . . . . **Drypetes**
35. Stamens mostly 3 or 6:
38. Styles 1–2, shortly reniform-flabellate . . . **Drypetes**
38. Styles 2–3, stout, clavate-capitate . . . **Neoroepera**
38. Styles 3–6, slender, sometimes coiled or very short, rarely flabellate:
39. Flowers dioecious; stamens free:
40. Tepals 5; stamens 3–5; fruit a small globose short-pedicelled drupe, with 6 pyrenes; leaves persistent **Securinega**
40. Tepals and stamens 4; fruit larger, dry, globose, often long-pedicelled, bursting irregularly; leaves deciduous **Margaritaria**
39. Flowers monoecious:
41. Fruit a normal dry capsule, dehiscing completely into 3–6 segments; top of ovary not excavated, without raised rim; stamens various:
42. Disk or glands present . . . . . **Phyllanthus**
42. Disk or glands absent . . . . . . **Sauropus**
41. Fruit somewhat fleshy, becoming dry, with incomplete dehiscence; top of ovary sometimes excavated, with raised rim; stamens connate; no true disk or glands:
43. Plant often blackening on drying; ♂ calyx turbinate, obconic, or hemispheric; styles short, erect **Breynia**
43. Plant not blackening; ♂ calyx sometimes disciform; styles sometimes ± circinate . . . . **Sauropus**
34. Flowers ± diffusely cymose or thyrsoid (see also 34 below):
44. ♂ tepals imbricate; plant mostly scandent; petiole biglandular at apex; stamens connate . . . . **Omphalea**
44. ♂ tepals valvate; erect shrubs or trees; stamens free:

45. Anthers 3–4-locellate; leaves never stipellate, but often glandular-granular beneath . . . . **Macaranga**
45. Anthers 2-locular; leaves sometimes stipellate, never glandular-granular . . . . **Alchornea**
34. ♂ flowers spicate or racemose:
[46. ♂ tepals imbricate . . . . **Antidesma**]
46. ♂ tepals valvate (see 46 below):
47. Anthers 4-locellate; capsule 2-locular (if rarely 3-locular, 1 loculus abortive); ♀ flowers and capsules borne on stiff elongate pedicels . . . . **Cleidion**
47. Anthers bilocular; capsules usually 3- (rarely 2-)locular, not borne on elongate pedicels:
48. Anthers with distinct, free, ± erect or spreading thecae:
49. Anther-thecae elongate, vermiform, twisted; no 'juxtastaminal glands' or purple sap . . . . **Acalypha**
49. Anther-thecae short, erect; filaments accompanied by short, often pilose, 'juxtastaminal glands' at base; capsules often with purple sap . . . . **Claoxylon**
48. Anthers with adnate or pendulous thecae:
50. Stamens 4–8; leaves sometimes stipellate **Alchornea**
50. Stamens numerous; leaves never stipellate . **Mallotus**
46. ♂ calyx much reduced, aestivation open or obscure:
51. Annual herb of sandy ground, sometimes perennating from woody rootstock; leaves linear to narrowly elliptic; capsule-lobes dorsally spinulose-dentate; seed caruncluate **Sebastiania**
51. Shrubs or trees with broader leaves; capsules unarmed; seed ecarunculate . . . . **Excoecaria**

Acalypha *L.*

Small trees or more commonly shrubs (occasionally scrambling) or nettle-like perennial or annual herbs, mostly monoecious. Leaves alternate, long- or short-petioled, crenate or dentate or subentire, palmatinerved or penninerved; stipules lanceolate or subulate or setaceous, sometimes minute. Inflorescences unisexual or bisexual, when bisexual the sexes very diversely arranged, the males commonly in slender dense-flowered spikes with 1–few females at the base, or the females alone in relatively short and less dense-flowered racemes, the bracts then often lobed or dentate and accrescent in fruit. Male flower (mostly minute): calyx closed in bud, splitting valvately into 4 segments; petals 0; disk 0; stamens mostly 8, free, anthers with free, narrow, twisted, vermiform loculi; pistillode 0. Female flower: sepals 3–5, shortly connate, imbricate; ovary 3- or 2-locular, loculi 1-ovulate, styles mostly conspicuous and laciniate. Fruit a small 3- or 2-locular capsule; seeds rounded, smooth, sometimes with a conspicuous hilum or caruncle.

1. Weedy annual to 50 cm tall, the ♀ flowers mostly 1–2 in each inflorescence, subtended by large ovate bracts 1–1·5 cm long usually concealing both ♂ and ♀ flowers . . . . A. AUSTRALIS

1. Perennial herbs or shrubs or small trees:
  2. Planted shrubs; ♀ inflorescences many-flowered; leaves up to 20 × 14 cm, often variously coloured (cf. also A. COMPACTA, with coppery leaves up to 7 × 2·5 cm, probably a sport) . . . . . A. WILKESIANA
  2. Native shrubs or small trees; ♀ inflorescences (or ♀ portions of bisexual inflorescences) few-flowered:
    3. Inflorescences unisexual, the ♀ long-peduncled; plant often ± spinose **A. capillipes**
    3. Inflorescences bisexual, the ♀ flowers shortly pedicelled:
      4. Nervation ± palmate; bract of ♀ flower up to 1·5 cm diam.; plant often villous, not or rarely spinose . . . . . . **A. nemorum**
      4. Nervation ± pinnate; bract of ♀ flower 4–5 mm diam.; plant glabrous, or puberulous when young, often spinose . . . **A. eremorum**

ACALYPHA AUSTRALIS *L.*, Sp. Pl.: 1004 (1753); Merr., Enum. Philipp. Fl. Pl. 2: 445 (1923); Pax & Hoffm. xvi: 35 (1924). Type: 'America meridionalis', *leg.*? (LINN).

A. gemina (Lour.) Spreng. var. *genuina* Muell. Arg. in Linnaea 34: 41 (1865) & in DC.: 866 (1866). Type: as for *A. australis* L.

A. indica var. *australis* F. M. Bailey in Queensl. Dep. Agric. Bull. 9: 16 (1891) & Queensl. Fl. 5: 1443 (1902); Domin: 886 [332] (1927), *in obs.* Type: Q., Walsh River, *Barclay-Millar* (BRI).

Q (CK, MO)—Native of E Asia (Manchuria and Japan to S China, Formosa, and N Philippines).

Introduced into the Botanic Gardens, Brisbane, 1930, 1942. A weedy annual herb, 10–50 cm tall, readily recognizable from the broad ovate bracts, 1–1·5 cm long, subtending the 1–2 female flowers of the inflorescence; male part of the inflorescence mostly slender and inconspicuous.

F. M. Bailey's *A. indica* var. *australis* appears to have been named without reference to *A. australis* L.

Acalypha capillipes *Muell. Arg.* in Linnaea 34: 40 (1865) & in DC.: 823 (1866); Benth.: 133 (1873); Moore & Betche: 77 (1893); Bailey: 1444 (1902); Pax & Hoffm. xvi: 173 (1924). Type: N.S.W., Clarence River, [*Beckler* 19] (MEL, K).

A. eremorum var. *capillipeda* Baill. in Adansonia 6: 317 (1866). Type: as for *A. capillipes*.

A. spinescens Benth. in Hook. Ic. Pl. 13: 72, t. 1291 (1879). Type: Celebes, *Riedel s.n.* (K).

Q (PC, MO); **NSW** (NC)—Celebes.

Erect virgate shrub with several branches from short trunks, to 3 m high, in dryish rain-forest up to 700 m.

Very closely related to *A. eremorum*, and perhaps, as Baillon thought, scarcely specifically distinct, but more glabrous, more spinescent, with the groups of male flowers more distantly arranged on the inflorescence, and the females 'apparently all solitary on filiform peduncles of 2–2·5 cm, with a normal orbicular bract of nearly 4 mm diameter, the capsule glabrous [or puberulous] and tridymous but the cocci not deeply separate as in the pedunculate fruits of *A. eremorum*, the styles with much more numerous capillary branches than in that species' (Bentham, *l.c.*).

ACALYPHA COMPACTA *Guilfoyle* ('hort. Brisbane') *ex C. T. White* in Gard. Chron. III, 94: 343, figs. 151, 152 (1933). Type: Q., Brisbane Botanic Garden, *C. T. White* (BRI).

Q (MO)—Cultivated shrub.

Shrub of 2–3 m; leaves of sterile shoots copper-coloured, 3–7 × 1–2·5 cm, petiole 1–2 cm, those of fertile reversion shoots up to 12 × 6 cm, marked with blood-red patches.

Planted in Brisbane Botanic Garden as an ornamental. Probably a sport of *A. wilkesiana* Muell. Arg.

Acalypha eremorum *Muell. Arg.* in Flora 47: 440 (1864) & in DC.: 863 (1866); Baill. in Adansonia 6: 317 (1866) (var. *sessilis* Baill.); Benth.: 132 (1873); Bailey: 1444 (1902); Pax & Hoffm. xvi: 114 (1924); Domin: 886 [332] (1927). Syntypes: Q., 'Bardikia' [Burdekin] River & Brisbane River, *F. Mueller* (G-DC, MEL).

A. eremorum var. *sessilis* Baill., *l.c.* (1866). Type as for *A. eremorum.*

Q (NK, LT, PC, MO)—Endemic.

Twiggy shrub to 2·5 m tall, in brigalow scrub or woodland forest or cleared rain-forest, on sandy soil over clay, up to 700 m.

Slender, with the lateral branchlets often terminating in spines. Leaves very small, 0·5–2·5 cm, oblong to obovate or suborbicular, glabrous, or puberulous when young, crenate, very short-petioled. Male inflorescence 1–2·5 cm, slender, rather dense-flowered. Female flower solitary, subsessile; bract 4–5 mm diam.; ovary trilocular, puberulous; styles with rather few capillary branches; capsule tricoccous, puberulous or glabrescent. Occasionally allomorphic ♀ flowers are produced on filiform pedicels 8–9 mm long, with a much reduced bract, the capsule deeply divided into 3 muricate cocci (cf. Radcliffe-Smith in Kew Bull. 28: 525–9 (1973) & 30: 676 (1975)).

Acalypha nemorum *F. Muell. ex Muell. Arg.* in Linnaea 34: 38 (1865) & in DC.: 858 (1866); Baill. in Adansonia 6: 318 (1866); Benth.: 132 (1873); Moore & Betche: 77 (1893); Bailey: 1443 (1902); Pax & Hoffm. xvi: 131 (1924); Beadle, Evans & Carolin: 252 (1972). Type: N.S.W., Clarence River, *F. Mueller* (G–C).

A. cunninghamii Muell. Arg. in Linnaea 34: 35 (1865) & in DC.: 861 (1866); Baill., l.c. (1866); Domin: 884 [330], fig. 151 (1927). Type: Q., Moreton Bay, *Cunningham* (G–DC).

Q (LT, BT, WB, MO); **NSW** (NC, CC)—Endemic.

Erect shrub to 1·5 m tall, common in semi-rain-forest or on rain-forest margins, or on rocky slopes in gully (Brisbane schist), or in Eucalypt forest on yellowish brown sandy loam or on gravelly clay, at unknown altitude.

Very variable: usually recognizable from its more robust habit, larger leaves and strong velutinous indumentum, but sometimes slender, with smaller leaves and very scanty short pubescence. The large fruiting bract, up to 1·5 cm diam., is generally a decisive character.

ACALYPHA WILKESIANA *Muell. Arg.*: 817 (1866); Pax & Hoffm. xvi: 153 (1924); Backer & Bakh. f.: 489 (1963); Airy Shaw in K.B. Add. Ser. IV: 24 (1975). Type: 'In insulis Fidji', 1838–42, *Wilkes* (G–DC).

Q (CK), Green Island, planted as an ornamental, and doubtless elsewhere.—Probably native of Polynesia.

Shrub of 2–3 m on sand cay.
Leaves variegated red and green, or mottled dark and light green.

Actephila *Bl.*

Trees or shrubs, mostly subglabrous. Leaves alternate, sometimes pseudo-verticillate, entire, long- or short-petioled, coriaceous or chartaceous, penninerved; stipules small, triangular or ovate. Flowers monoecious, fascicled, axillary and extra-axillary (rarely borne on short, densely and minutely bracteate brachyblasts). Male flower: sepals 5–6, imbricate, very shortly connate; petals 5–6, shorter than sepals, ± unguiculate-spathulate, sometimes 2–3 or reduced or 0; disk extra-staminal, flat, entire or sometimes lobed, lobes episepalous; stamens (3–)5(–6), inserted on the disk, free or shortly connate; pistillode trifid. Female flower: sepals and petals ± as male; disk flat or cupular; ovary 3-locular, loculi biovulate, styles free or ± connate, entire or bifid. Capsule tricoccous, long- or sometimes short-pedicelled; seeds exarillate, cotyledons thick and fleshy, sometimes plicate, endosperm 0.

1. Leaves subsessile, cordate, oblong, up to 7·5 × 3 cm . **A. sessilifolia**
1. Leaves clearly petioled, cuneate or rounded but not cordate at the base:
 2. Leaves large, up to 30 × 13 cm, shortly puberulous beneath, olive-green when dry; stipules densely adpressed-pubescent; flowers puberulous . . . . **A. foetida**
 2. Leaves not exceeding 18 × 10 cm; plant practically glabrous:
  3. Petiole slender, up to 7·5 cm long; lamina mostly ovate, smooth and shining, bright green when dry . . . . **A. petiolaris**
  3. Petiole not exceeding 2·5 cm:
   4. Leaves broadly elliptic, up to 15 × 7·5 cm, coriaceous, with a very rough surface, brown when dry; petiole 1–2·5 cm long **A. latifolia**
   4. Leaves mostly cuneate-obovate to cuneate-oblanceolate, very variable in size, less coriaceous, smooth, mostly green when dry, sometimes pseudo-verticillate; petiole occasionally almost obsolete **A. lindleyi**

Actephila foetida *Domin*: 869 [315] (1927); Airy Shaw in K.B. 31: 363 (1976) & in Muelleria 4: 217 (1980). Type: Q., Harvey's Creek, Dec. 1909, *Domin* s.n. (PR).

Q (CK)—Endemic.

Shrub of 1–2 m (*Hyland*) or 'moderate sized tree' (*Domin*), in rain-forest at 80 m.

Branchlets puberulous, strongly lenticellate. Leaves broadly elliptic, up to 30 × 13 cm, rounded at the base, obtuse to shortly caudate-acuminate, olive-green when dry, very shortly puberulous beneath, glabrous above; primary nerves 12–13 pairs; petiole up to 4 cm long; stipules triangular, 1–2 mm long, densely adpressedly ochraceous-pubescent. Flowers in mixed axillary fascicles, puberulous; pedicels of males ± 4 mm, of females ± 8 mm long; sepals 5, obtuse, ± 2 mm long; petals 0; disk conspicuous; stamens 5; ovary glabrous; styles 3, simple, 1 mm long, stigmas subcapitate. Capsule 5–6 mm long; pedicel 9–10 mm long, rigid; seeds not seen.

This scarce species has only been collected on three occasions (by Sayer, Domin and Hyland), and only in the neighbourhood of Harvey's Creek. The large leaves, olive-green when dry, and the short pale puberulence of all parts except the upper leaf-surface, distinguish the plant clearly from the remaining species.

Actephila latifolia *Benth.*: 89 (1873); Bailey: 1414 (1902); Pax & Hoffm. xv: 194 (1922). Lectotype (present designation): Q., Cape York, *Daemel* (K) (syntype: Rockingham Bay, *Dallachy*).

[*A. lindleyi* sec. Airy Shaw in K.B. 25: 496–8 (1971), *pro parte*, vix *Lithoxylon lindleyi* Steud.]

Q (CK, NK)—Key Islands (*testibus* Pax & Hoffmann, *l.c.*).

No field information.

At present only known from the two syntype gatherings of Daemel and Dallachy, from Cape York and Rockingham Bay respectively. As far as it is possible to judge from this limited material, *A. latifolia* differs from the closely related *A. lindleyi* in the coriaceous, broadly elliptic, longer-petioled leaves, rounded or very broadly cuneate at the base and up to 15 × 7·5 cm in size, and in the absence of petals in the flowers. My reduction of this species to *A. lindleyi* in 1971 was possibly premature, but in view of the extraordinary variability of *A. lindleyi* it is by no means impossible that further material of *A. latifolia* might eventually prove the reduction to have been justified.

Actephila lindleyi (*Steud.*) *Airy Shaw* in K.B. 25: 496 (1971), q.v. for discussion, & in K.B. Add. Ser. VIII: 23 (1980). Type: Cult. in Hort. Kew., 1813–1821 (K).

[*Securinega nitida* sec. Ait. f., Hort. Kew, ed. 2, 5: 383 (1813); Sm. in Rees, Cyclop. 32 (1819), *p.p.*; Lindl., Collect. Bot.: t. 9 (1821); *non* Willd. (1806); cf. Juss., Euphorb. Gen. Tent.: 14 (1824), *in obs.*]

Lithoxylon lindleyi Steud., Nomencl. Bot., ed. 2, 2: 57 (1841), *pro nom. nov.*

L. nitidum [Lindl.] Baill., Ét. Gén. Euphorb.: 590 (1858); Muell. Arg.: 232 (1866); Nadeaud, Enum. Pl. Tahiti: 72 (1873); *nom. illegit.*, *superfl.*

L. grandifolium Muell. Arg. in Linnaea 34: 65 (1865) & in DC.: 232 (1866). Type: N.S.W., Clarence River, *F. Mueller* (G–C).

Actephila grandifolia (Muell. Arg.) Baill. in Adansonia 6: 330, 360, t. 10 (1866); Benth.: 89 (1873); Moore & Betche: 73 (1893); Bailey: 1413 (1902); Pax & Hoffm. xv: 193 (1922); Anderson: 218, 377 (1968); Beadle, Evans & Carolin: 252 (1972).

?*A. mooriana* Baill. in Adansonia 6: 330 (*nomen*), 366 (1866); Benth.: 89 (1873); Moore & Betche: 73 (1893); Bailey: 1414 (1902); Pax & Hoffm. xv: 193 (1922); Anderson: 217, 377 (1968). Type: N.S.W., Mt Lindsay, Richmond River, 1861, *C. Moore* 211 (P, MEL).

A. nitida (Baill.) 'Benth. & Hook. f.' ex Drake del Castillo, Illustr. Ins. Maris Pacif.: 286 (1892) & Fl. Polynésie Franç.: 177 (1893); Pax & Hoffm. xv: 194 (1922); sed cf. Benth. & Hook. f., Gen. Pl. 3: 270 (1880); *nom. illegit., superfl.*

Q (CK); **NSW** (NC, CC)—Moluccas, Lesser Sunda Is., New Guinea, Solomons.

Shrub or slender tree to 4 m tall, in undergrowth of rain-forest or gallery forest at 60–80 m.

Closely related to *A. latifolia*, differing principally in its rather thinner, shorter-petioled, mostly cuneate-oblanceolate leaves and in the presence of small petals in the flowers. Like other species of the genus, e.g. *A. excelsa* (Dalz.) Muell. Arg. and *A. javanica* Miq., it is an extraordinarily variable plant, and there is need of much more material before the limits of its variation can begin to be understood.

Actephila petiolaris *Benth.*: 89 (1873); Bailey: 1414 (1902); Pax & Hoffm. xv: 194 (1922); Airy Shaw in Muelleria 4: 217 (1980). Type: Q., Rockingham Bay, *Dallachy* (K).

Q (CK)—Endemic.

Shrub or slender tree of 5–6 m, in rain-forest at 100–160 m. Fig. 1A.

Quite glabrous, or branchlets and flowers sometimes minutely puberulous; leaves ovate or obovate, 7–18 × 4–9·5 cm, broadly rounded or sometimes broadly cuneate at base, cuspidate or shortly acuminate at the apex, smooth and somewhat shining, green when dry or plumbeous above; petiole 2–7·5 cm long, slender; stipules and perulae triangular, 1–2 mm long, acute, convex, pale; flowers sparse, in few-flowered axillary fascicles. Male flowers with 3–5 stamens; female flower and fruit unknown.

The glabrous, ovate leaves, drying bright green, on usually long slender petioles, are characteristic of this species. Known only from the type collection for over 100 years, until re-collected by Hyland in 1974 in State Forest Reserve 675, East Mulgrave Logging Area, in the SW foothills of Mt Bellenden-Ker.

Actephila sessilifolia *Benth.*: 90 (1873); Bailey: 1414 (1902); Pax & Hoffm. xv: 194 (1922); Airy Shaw in Muelleria 4: 217 (1980). Type: Q., 'Caves mountains, five miles west of Morinisi', *Thozet* (K).

Q (NK, PC)—Endemic.

Shrub of 1–2 m, in rain-forest on soil derived from granite, at 250–915 m. Fig. 1B.

Branchlets slender, with rough, scaberulous, minutely puberulous bark; leaves oblong-elliptic or narrowly obovate, 2·5–7·5 × 1·5–3 cm, shortly cordate at the base, obtuse or rounded at the apex, with a 'shagreened' or slightly roughened surface, pinkish beneath when dry and plumbeous above, glabrous; petioles 1–2 mm long or almost obsolete, glabrous; stipules triangular, 1–2 mm long, glabrous. Male flowers unknown. Female flowers (according to Bentham): pedicels slender, calyx deeply 6-lobed, about 4 mm diam., with very small petals; styles 3, united at base, short, spreading, undivided. Capsule depressed-subglobose, about 12 mm diam., granular-punctate, shortly beaked, on a slender pedicel nearly 3 cm long. Seeds not seen.

The subsessile, cordate, oblong leaves make this an outstanding species in the genus. Hopefully the male flowers will be forthcoming in due course, but the flowers of *Actephila* appear to furnish few characters of taxonomic value. The three so far known localities for the species (Seaview Range, Mt Dryander, and Caves Mountain) are widely separated.

Adriana *Gaudich.*

Low, bushy shrubs, glabrous or stellate-pubescent, dioecious. Leaves alternate or opposite, simple or 3(–5)-lobed, petiolate or subsessile, mostly rather coarsely toothed; stipules small, variable, discoid or shortly cylindric and umbilicate at the apex as though hydathodal. Inflorescences terminal or leaf-opposed, apparently spicate (the ♂ elongate and interrupted, the ♀ congested and few-flowered), but actually very narrowly thyrsoid, each ♂ bract subtending a cymule of 3–5 flowers, the central flowers of all the cymules expanding more or less simultaneously, followed by the laterals (cf. *Verbascum*, *Cordia*, etc.). Male flower: sepals 3–5, valvate; petals and disk 0; stamens very numerous, on a convex or subglobose receptacle, the filaments very short, the anthers curvilinear, erect, basifixed, the connective produced into a short, liguliform, cellular-papillose appendage, amber-coloured and glistening when dry; pistillode 0. Female flower: sepals 3 + 3 or 4 + 4, biseriate, imbricate, narrowly elliptic or ovate, acute; ovary 3-locular, ± muricate, stellate-pubescent or sericeous; styles elongate, free or shortly connate, deeply bifid, segments linear, coarsely papillose, crimson. Capsule tricoccous, dehiscent; seeds ovoid, shortly carunculate.

The inter-specific taxonomy of *Adriana* has caused botanists many problems from F. Mueller to the present day.

1. Leaves opposite, sessile or subsessile:
  2. Leaves glabrous; styles shortly connate . . . **A. quadripartita**
  2. Leaves hoary stellate-tomentellous beneath; styles free . **A. klotzschii**
1. Leaves alternate, petiolate:
  3. Styles shortly connate below; leaves oblong-lanceolate, rounded at apex, obtusely lobulate, never glabrous; side-lobes short and broad or none; veins scarcely impressed . . . . **A. hookeri**
  3. Styles free; leaves acutely or obtusely dentate, sometimes glabrous; side-lobes when present longer, often acute; nerves finely incised above:

FIG. 1. *Actephila petiolaris:* **A1** habit × ⅓, from *Hyland* 9249. *A. sessilifolia:* **B1** habit × ⅓, from Herb. Mueller 1872; **B2** fruit × 1⅓. *Cleistanthus xerophilus:* **C1** habit × ⅓, from *Hyland* 8274; **C2** leaf × ⅓, from *Hyland* 8419. *Claoxylon tenerifolium:* **D1** habit × 2⅓, from *Hyland* 7715; **D2** flower × 6, from *Hyland* 7715; **D3** fruit × 2 from *Hyland* 7743.

4. Leaves always 3-lobed; lobes either ± parallel-sided or broadest near middle . . . . **A. tomentosa**
4. Leaves often simple, unlobed; lobes when present rarely parallel-sided, broadest nearer base . . . . **A. glabrata**

Adriana glabrata *Gaudich.* in Ann. Sci. Nat. I, 5: 223 (1825) & in Bot. Voy. Freyc.: 487 (1826); Pax & Hoffm. ii: 18 (1910); Black, Fl. S. Austr. 2: 352 (1924); Willis, Handb. Pl. Vict. 2: 347 (1972). Type: N.S.W., 'Habitat in Novâ-Hollandiâ (Orientali), Nouvelle Galles du Sud (Hawkesbury River)', *D'Urville & Lesson* (P).

A. acerifolia var. *glabrata* (Gaudich.) Benth.: 134 (1873); Bailey: 1441 (1902).

var. **subglabra** (*Baill.*) *Airy Shaw* comb. nov.

A. gaudichaudi var. *subglabra* Baill. in Adansonia 6: 312 (1866). Syntypes: N.S.W., Hawkesbury River, *Baudin* (P); *Sieber* 569 (P).
A. acerifolia Hook. in Mitch., Journ. Exped. Trop. Austr.: 371 (1848); Baill. in Adansonia 6: 312 (1866); Ewart & Davies: 168 (1917); Burbidge & Gray, Fl. A.C.T.: 248, fig. 244 (1970). Syntypes: N.S.W., Lachlan River, *Cunningham* (K); Maranoa River, *Mitchell* (K).
[*A. tomentosa* sec. F. Muell. in Trans. Bot. Soc. Edinb. 7: 482 (1863), *pro parte*, non Gaudich.]
A. acerifolia var. *genuina* Muell. Arg.: 890 (1866).
?*A. acerifolia* var. *lessoni* Baill. in Adansonia 6: 313 (1866). Syntypes: N.S.W., 'Nouv.-Holl. Orient.', *D'Urville, Lesson* (P).
A. glabrata var. *acerifolia* (Hook.) Pax, *l.c.* (1910); Beadle, Evans & Carolin: 253 (1972).

WA (K); **NT** (DG); **Q** (MO); **NSW**; **V** (E)—Endemic.

Leaves variously pubescent.
Straggling or rounded bushy shrub 1–3 m high and through, on creek banks in deep sandy loam, or in sand and silt deposits in river gorge, or on grey limestone soil, up to 450 m.

var. **glabrata**:

A. heterophylla Hook. in Mitch., Journ. Exped. Trop. Austr.: 124 (1848). Type: Q., Maranoa, Balonne River, Camp VII, *Mitchell* (K).
Trachycaryon cunninghamii var. *glabrum* F. Muell. in Trans. Phil. Soc. Victoria 1: 15 (1855) & in Hook. Kew Journ. Bot. 8: 124 (1856). Type: V., between granite rocks and on the sandy banks of the Snowy River, *F. Mueller* (MEL).
[*Adriana tomentosa* sec. F. Muell. in Trans. Bot. Soc. Edinb. 7: 482 (1863), *pro parte*, non Gaudich.]
A. glabrata var. *heterophylla* (Hook.) Muell. Arg.: 891 (1866); Pax & Hoffm. ii: 18 (1910); Willis, Handb. Pl. Vict. 2: 347 (1972).
A. acerifolia var. *glabrata* (Gaudich.) Benth.: 134 (1873); Bailey: 1441 (1902).

Leaves glabrous or almost so on both surfaces.

var. **cunninghamii** (*F. Muell.*) *Muell. Arg.*: 890 (1866); Pax ii: 18 (1910); Willis, Handb. Pl. Vict. 2: 347 (1972). Type: V., 'between granite rocks and on the sandy banks of the Snowy River', *F. Mueller* (MEL).

Trachycaryon cunninghamii F. Muell. in Trans. Phil. Soc. Victoria 1: 15 (1855).
T. cunninghamii var. *tomentosum* F. Muell. in Trans. Phil. Soc. Vict. *l.c.* (1855) & in Hook. Journ. Bot. & Kew Garden Misc. 8: 209 (1856). Type: V., between granite rocks and on the sandy banks of the Snowy River, *F. Mueller* (MEL). [Non *Adriana tomentosa* Gaudich.]
[*Adriana tomentosa* sec. F. Muell. in Trans. Bot. Soc. Edinb. 7: 482 (1863), *pro parte*, non Gaudich.]

Leaves almost sessile.

Adriana hookeri (*F. Muell.*) *Muell. Arg.*: 891 (1866); Pax ii: 20 (1910); Black, Fl. S. Austr. ed. 2, 2: 514 (1948); Willis, Handb. Pl. Vict. 2: 347 (1972).

var. **hookeri:**

Trachycaryon hookeri F. Muell. in Trans. Phil. Soc. Victoria 1: 16 (1855). Type: V., 'on sand-ridges along the Murray, towards the junction of the Darling and the Murrumbidgee', *F. Mueller* (MEL).
T. hookeri var. *velutinum* F. Muell. in Hook. Journ. Bot. & Kew Garden Misc. 8: 210 (1856). Type: V.,' on sand-ridges along the Murray, towards the junction of the Darling and Murrumbidgee', *F. Mueller* (MEL).

WA (Er–Ca); **SA** (N & W); **NT** (CA); **V** (NW); **NSW** (N, C & SC, N, C & ST, N, C & SWS, WP)—Endemic.

Plant finely pubescent.

var. **glabriuscula** (*F. Muell.*) *Muell. Arg.*: 891 (1866); Pax ii: 20 (1910).

Trachycaryon hookeri var. *glabriusculum* F. Muell. in Hook. Journ. Bot. & Kew Gard. Misc. 8: 210 (1856). Type: V., 'on sand-ridges along the Murray, towards the junction of the Darling and the Murrumbidgee', *F. Mueller* (MEL).

Plant almost glabrous.

Densely spreading shrub 1–1·5 m high and wide, in mallee on pale sandy soil or on red sandy soil or in inter-dune flats at low altitude.

This form is probably scarcely worth recognizing as a distinct taxon.

Adriana klotzschii (*F. Muell.*) *Muell. Arg.*: 892 (1866); Benth.: 135 (1873); Pax ii: 21 (1910); Black, Fl. S. Austr., ed. 2, 2: 514 (1948); Willis, Handb. Pl. Vict. 2: 347 (1972). Type: V., 'on sandhills near Corner Inlet, and in various localities in South Australia', *F. Mueller* (MEL).

Trachycaryon klotzschii F. Muell. in Trans. Phil. Soc. Victoria 1: 15 (1855) & in Hook. Journ. Bot. & Kew Gard. Misc. 8: 209 (1856).
[*Adriana billardieri* sec. F. Muell. in Trans. Bot. Soc. Edinb. 7: 482 (1863), *pro parte*, non Baill.]

W (SE); **SA** (S & SE); **V** (W, S)—Endemic.

Shrub of 1–2 m, by lake near sea or in coastal sand-dunes or in red sand by creek at low altitude.

Adriana quadripartita (*Labill.*) *Gaudich.*, Bot. Voy. Freyc.: 489 (1826); Muell. Arg.: 892 (1866); Benth.: 135 (1873); Pax ii: 20 (1910); Gardner: 72 (1931); Blackall & Grieve, W. Austr. Wildfl. 1: 265 (1954, 1974); Cochrane, Fuhrer, Rotherham & Willis, Flowers & Pl. Vict.: 100, fig. 289 (1968); Willis, Handb. Pl. Vict. 2: 347 (1972). Type: W.A., 'habitat in capite Van-Diemen', *Labillardière* (P).

Croton quadripartitus Labill., Nov. Holl. Pl. Nov. 2: 73, t. 223 (1807).
Meialisa australis Rafin., Sylva Tellur.: 64 (1838), *nom. illegit.*, superfl. Type: W.A., *Labillardière* (P).
Trachycaryon labillardierei Klotzsch in Lehm., Pl. Preiss. 1: 175 (1844–5), *nom. illegit. superfl.* Type: W.A., *Labillardière* (P).
Adriania [sic] *billardieri* Klotzsch ex Baill., Ét. Gén. Euphorb.: 406, t. 2, figs. 19–22 (1858), *nom. illegit.*, *superfl.* Type: W.A., *Labillardière* (P).

WA (SW); **SA**; **V** (S)—Endemic.

Shrub of 1·5 m, in deep sand on river bank near sea at low altitude.

Adriana tomentosa *Gaudich.* in Ann. Sci. Nat. I, 5: 223 (1825) & in Bot. Voy. Freyc.: 487, t. 116 (1826); Baill., Ét. Gén. Euphorb., Atlas: 18, fig. 12 (1858); F. Muell. in Trans. Bot. Soc. Edinb. 7: 482 (1863), *pro parte*; Muell. Arg.: 891 (1866); Benth.: 134 (1873); Moore & Betche: 77 (1893); Pax ii: 18 (1910); Ewart & Davies: 168 (1917); Gardner: 72 (1931); Blackall & Grieve, W. Austr. Wildfl. 1: 265 (1954, 1974). Type: W.A., 'Habitat in Novâ Hollandiâ (Orientali [*sic*! *scil.* Occidentali]), Baie des Chiens-Marins [Sharks Bay]', *D'Urville & Lesson* (P).

A. acerifolia var. *puberula* Muell. Arg.: 891 (1866). Type: W.A., 'ad Murchison River', *F. Mueller* (G–DC, MEL).
A. gaudichaudi Baill. in Adansonia 6: 312 (1866), var. *genuina* Baill., *l.c.* Syntypes: W.A., Baie des Chiens marins [Sharks Bay], *Baudin* (P); Voy. Uranie, *Gaudichaud* (P).
A. gaudichaudi var. *thomasiaefolia* Baill., *l.c.* (1866). Syntypes: W.A., Baie des Chiens marins, *Gaudichaud* 20, 21 (P).

WA (N fo, E ash)—Endemic.

Erect or straggling shrub to 1 m high, in watercourses, in dry sand-dunes or in heavy red sandy soil on flat between dunes at low altitude.

Adriana tomentosa is not *specifically* distinct from *A. glabrata*, but is conveniently maintained because of the geographical disjunction.

Alchornea *Sw.*

Trees or shrubs, monoecious or dioecious, glabrous or ± pubescent. Leaves alternate, ± crenate or dentate, chartaceous or membranaceous, either spathulate-oblanceolate, penninerved and short-petioled, or elliptic-ovate, palmatinerved and long-petioled (and then bistipellate), sometimes bearing a few basal macular glands. Inflorescences of simple or compound racemes or spikes, bracts 1–several-flowered, sometimes biglandular. Male flower: calyx closed in bud, splitting valvately into 2–5 segments; petals 0; disk 0; stamens 3–8, filaments shortly connate into a small disc, anthers shortly oblong, introrse; pistillode 0. Female flower: sepals 4–8, free, sometimes 1–4-glandular; ovary (2–)3-locular, loculi l-ovulate, styles (2–)3, free or very shortly connate, simple, often elongate. Capsule (2–)3-locular, lobed or unlobed, smooth or muricate.

1. Inflorescences lateral; leaves chartaceous to coriaceous, mostly under 10 cm long, holly-like, with spine-tipped lobes; styles variable **A. ilicifolia**
1. Inflorescences terminal; leaves chartaceous to herbaceous; styles ± subulate:
  2. Leaves similar to those of *A. ilicifolia* in size and general outline, but scarcely lobed, and teeth merely acute, not spinescent **A. thozetiana**
  2. Leaves spathulate or cuneate-oblanceolate, contracted to a narrowly cordate base, up to 22 cm long . . . . **A. rugosa**

Alchornea ilicifolia (*J. Smith*) *Muell. Arg.* in Linnaea 34: 170 (1865) & in DC.: 906 (1866); Benth.: 136 (1873); Moore & Betche: 78 (1893); Bailey: 1445 (1902). Type: Hort. Kew., 1829–1839, *J. Smith* (K).

Caelebogyne ilicifolia J. Smith in Trans. Linn. Soc. 18: 512, t. 36 (1841); Baill., Ét. Gén. Euphorb., Atlas: t. 8, figs. 32–36 (1858); Pax & Hoffm. vii: 256, figs. 38, 39 (1914); Anderson: 218 (1968); Beadle, Evans & Carolin: 252 (1972).

Cladodes ilicifolia (J. Smith) Baill. in Adansonia 6: 321 (1866).

Q (CK, NK, PC); **NSW** (NC, CC)—Endemic.

Shrub or small tree to 7 m high, in dryish or transitional rain-forest on sandy loam up to 120 m.

Leaves holly-like, 2–7(–14) × 2·5(–9) cm, broadly elliptic or rhomboid or obovate, cuneate at base, with 2–4 pairs of broadly triangular spine-tipped lateral lobes, harshly chartaceous to coriaceous, glabrous, prominently nerved. Male and female inflorescences short, lateral, little branched. Ovary glabrous or minutely puberulous; styles short, very variable, mostly acutely toothed or lacerate toward the apex, but sometimes simply acute or else flabellate with a broad subentire apex. Capsule virtually glabrous.

Alchornea rugosa (*Lour.*) *Muell. Arg.* in Linnaea 34: 170 (1865) & in DC.: 905 (1866); J. J. Sm.: 467 (1910); Pax & Hoffm. vii: 243 (1914); Airy Shaw in K.B. 26: 211 (1971) & 31: 393 (1976) & K.B. Add. Ser. IV: 28 (1975). Type: 'in sylvis Cochinchinae', *Loureiro* (BM).

Cladodes rugosa Lour., Fl. Cochinch.: 574 (1790).

Conceveibum javanense Bl., Bijdr.: 614 (1825). Type: Java, 'Bantam, Buitenzorg, etc.', *leg.*? (BO).

Aparisthmium javense Endl. ex Hassk., Cat. Pl. Hort. Bogor. Alter: 235 (1844), *nom. illegit.*, *superfl.* Type: as for *Conceveibum javanense* Bl.

Alchornea javensis (Endl. ex Hassk.) Muell. Arg. in Linnaea 34: 170 (1865) & in DC.: 905 (1866).
A. javanensis (Bl.) Backer & Bakh. f.: 485 (1963).

Q (CK, NK)—Nicobar Is., SE Asia and throughout Malesia to New Guinea.

Shrub or small tree to 4 m tall, in rain-forest or dry rain-forest, or on edge of complex notophyll vine-forest, on red soils derived from metamorphic rocks, up to 600 m.

Leaves spathulate or cuneate-oblanceolate, up to 22 × 8 cm, narrowly cordate at base, rounded and shortly, obtusely and abruptly cuspidate-acuminate at apex, acutely or obtusely repand-dentate; petiole 5–10 mm; young parts puberulous, later glabrescent. Inflorescences terminal, the male laxly branched, with glomerules of small sessile glabrous flowers, the female simply racemose, with an ochraceous-sericeous ovary and three spreading subulate styles. Capsule tricoccous, minutely puberulous.

Alchornea thozetiana (*Baill.*) *Benth.*: 137 (1873); Bailey: 1445 (1902). Type: Q., 'near Rockhampton', *Thozet* (MEL).

Cladodes thozetiana Baill. in Adansonia 6: 321 (1866).
Caelebogyne thozetiana (Baill.) Pax & Hoffm. vii: 257 (1914).

var. **thozetiana**

Q (SK, PC)—Endemic.

Shrub to 1·5 m tall, on rain-forest margins.

Almost exactly intermediate between *A. ilicifolia* and *A. rugosa*. Leaves with the general shape and size of the former, but with the chartaceous or firmly herbaceous texture of the latter; not or scarcely lobed, but with about 6 acute but scarcely pungent teeth on either side, rarely subentire. Inflorescences terminal as in *A. rugosa*; styles shortly subulate, shortly connate, entire or shortly toothed at the apex; capsule minutely puberulous.

var. **longifolia** *Benth.*: 137 (1873); Bailey: 1445 (1902). Type: Q., Rockingham Bay, *Dallachy* (K).

Q (CK, NK)—Endemic.

Shrub to 2 m high, in dry rain-forest at 500 m.

Leaves elliptic or cuneate-obovate, up to 10 × 4·5 cm, with 9–12 obtuse or acute or uncinate teeth on each side, approaching those of *A. rugosa*.

Aleurites *J. R. & G. Forst.*

Tall tree (to 36 m). Leaves alternate, simple, ovate, rhomboid or sometimes palmately 3–5-lobed, palminerved, thinly or densely stellate-pubescent, long-petioled, petiole biglandular at apex; stipules minute or obsolete. Flowers monoecious. Inflorescences thyrsoid, terminal and axillary, mostly ♂♀, the male flowers greatly outnumbering the females, the latter terminating the ultimate branchlets of the cymes. Male flower: calyx closed in bud, 2–3-lobed at anthesis; petals 5, free, conspicuous, spreading; disk glands 5; stamens 10–20, 3–4-seriate, the outer free, the inner connate; pistillode 0. Female flower larger than male; calyx and petals as male, but petals narrower; disk-glands 5; ovary ovoid, 2–3-locular, tomentose; styles 2–3, deeply bipartite. Fruit large, drupaceous, up to 6 × 5 cm, indehiscent; seeds compressed-globose, up to 3 × 3 cm.

Aleurites moluccana (*L.*) *Willd.*, Sp. Pl. 4: 590 (1805); Muell. Arg.: 723 (1866); Baill. in Adansonia 6: 297 (1866); [Benth.: 128 (1873); Bailey: 1434 (1902);] Pax in Engl. IV. 147 (Heft 42): 129 (1910); J. J. Sm.: 551 (1910); Backer & Bakh. f.: 478 (1963); Airy Shaw in K.B. 20: 393 (1966) & 26: 213 (1971) & K.B. Add. Ser. IV: 29 (1975); Hyland: 22 (1971). Type: 'Habitat in Moluccis, Zeylona'; cult. Ceylon, *Hermann* (BM).

Jatropha moluccana L., Sp. Pl.: 1006 (1753).
Aleurites triloba J. R. & G. Forst., Char. Gen. Pl.: 111, t. 56 (1776). Type: South Sea Islands, *Forster* (whereabouts uncertain).
A. ambinux Pers., Synops. 2: 587 (1807); Baill., Ét. Gén. Euphorb.: 347, Atlas: t. 11, figs. 19, 20, t. 12, figs. 1–15 (1858), *nom. illegit. superfl.* Type: as for *Jatropha moluccana* L.

var. **moluccana**

Q (CK)—India and China to Polynesia and New Zealand.

Tree to 30 m high, in rain-forest or gallery rain-forest or dry rain-forest at 50–900 m.

Indumentum thin, evanescent; leaves relatively narrow, not or rarely cordate; ovary and fruit bilocular.

Var. *moluccana* occurs from the Cape York Peninsula south to lat. 16° 15′ S.

var. **rockinghamensis** *Baill.* in Adansonia 6: 297 (1866); Airy Shaw in Muelleria 4: 230 (1980), *q.v.* Type: Q., Rockingham Bay, *Dallachy* (P, MEL).

[*A. moluccana* sec. Benth.: 128 (1873), Bailey: 1434 (1902), quoad loc. 'Rockingham Bay'.]
?*A. moluccana* var. *floccosa* Airy Shaw in K.B. 20: 26 (1966). Type: Northeast New Guinea, Wau, *Havel* NGF 9169 (K).

Q (CK, NK)—? New Guinea.

Tree 7–30 m high in rain-forest or fringing forest (sometimes on sandy loam), or in dry rain-forest, at 15–900 m.

Indumentum evident, subfloccose; leaves broader, mostly cordate; ovary and fruit 3(–4)-locular.

The area of var. *rockinghamensis* extends from the latitude of the Daintree River (16° 10′ S) south to the region of Rockingham Bay.

Austrobuxus *Miq.*

Monoecious or dioecious trees or shrubs. Leaves simple, opposite or alternate, ± coriaceous, entire or rarely crenulate; stipules mostly obsolete, or occasionally minute, ovate. Inflorescences unisexual or bisexual, mostly cymose. Male flower: sepals 4–6, suborbicular, much imbricate; receptacle convex; stamens 2–25, arising from pits in the receptacle; pistillode 0. Female flower: sepals 4–6, ovate, acute; disk shortly cupular or obsolete; ovary 2–3-locular, loculi bi-ovulate or (in two New Caledonian species) uni-ovulate; stigmas 2–3, subsessile, cordate-ovate or transversely reniform, mutually adpressed or ± distant. Fruit a 2–3-locular capsule or drupe; seeds conspicuously arillate or with a small caruncle or ecarunculate.

1. Leaves elliptic or narrowly cuneate-obovate, entire; ovary and capsule 3–4-locular . . . . **A. nitidus**
1. Leaves lanceolate, crenulate-dentate; ovary and capsule bilocular **A. swainii**

Austrobuxus nitidus *Miq.*, Fl. Ind. Bat., Suppl.: 445 (1861); Airy Shaw in K.B. 25: 506 (1971) & 29: 309 (1974) & K.B. Add. Ser. IV: 43 (1975). Type: 'Sumatra occid, in littore prope Siboga,' *Teijsmann* (BO).

Choriophyllum malayanum Benth. in Hook. Ic. Pl. 13: 62, t. 1280 (1879). Lectotype (present designation): Borneo, Sarawak, *Beccari* 3329 (the only specimen at K written up in Bentham's handwriting). Original syntypes: Singapore, *Wallich* 7975; Malacca, *Griffith*, *Maingay*; Penang, *Maingay*; Borneo (Sarawak), *Beccari* 3270, 3305, 3329, 3344 (all K).
Longetia malayana (Benth.) Pax & Hoffm. xv: 291 (1922).
L. nitida (Miq.) v. Steenis in Blumea 12: 362 (1964) & 15: 155 (1967).
L. sp. (= RFK/1255), Hyland: 72 (1971).

Q (CK)—Malaya, Sumatra, Borneo.

Tree 20–30 m high in rain-forest at 700–1200 m.

Leaves opposite, elliptic or narrowly cuneate-obovate, 3–9 × 1–3 cm (up to 11 × 4·5 cm in coppice shoots), cuneate at base, obtuse or rounded at apex, margin entire or minutely glandular-indentate, reflexed or revolute, stiffly coriaceous, glabrous, reticulate-veined, somewhat glossy; petiole 5–8(–10) mm long; young parts cinnamomeous-pubescent. Inflorescences cymose, axillary, 1–2 cm long, the male few-flowered, minutely grey-puberulous, the female sometimes 1-flowered, cinnamomeous-pubescent. Male flower with 4–5 sepals and stamens; female flower with 2 + 2 sepals, a fimbriate shortly cupular disk, and an ovoid pubescent 3-locular ovary with a subsessile 3-lobed stigma. Capsule broadly obovoid, 1·5–2 cm long.

Austrobuxus swainii (*de Beuzev. & C. T. White*) *Airy Shaw* in K.B. 25: 508 (1971).

Longetia swainii de Beuzev. & C. T. White in Proc. Linn. Soc. N.S.W. 71: 236, fig. 1 (1947); Anderson: 220 (1968); Francis: 230 (1970). Type: N.S.W., East Dorrigo, 1944, *Miss Rosling* (NSW).

NSW (NC)—Endemic.

Tree to 40 m tall in rain-forest at unknown altitude.

Distinguished from *A. nitida* chiefly by its lanceolate, sinuate-crenulate leaves and bilocular ovary and capsule.

Baloghia *Endl.*

Shrubs or small trees, dioecious or monoecious. Leaves alternate or opposite, entire, coriaceous, mostly glabrous; stipules obsolete. Inflorescences racemose, lateral or terminal, uni- or bi-sexual (then lower flowers female). Male flower: sepals (4–)5(–6), shortly connate, imbricate; petals (4–)5(–6), free, obovate, elliptic, spathulate or oblong, mostly exceeding the sepals; disk annular or consisting of small alternipetalous glands or obsolete; stamens 10–100 (mostly ± 40–50), arising from a convex receptacle, filaments shortly connate or ± free, rather thick, anthers dorsifixed, extrorse, thecae oblong, parallel; pistillode 0. Female flower: sepals and petals ± as male; ovary 3(–4)-locular, loculi 1-ovulate: styles free or very shortly connate, deeply bifid, the branches simple or multifid. Capsule tricoccous; seeds oblong or globose, sometimes with a small caruncle.

1. Leaves opposite; inflorescence terminal; margins and tips of calyx-lobes frequently white-puberulous . . . . **B. lucida**
1. Leaves alternate; inflorescence lateral, mostly long-peduncled; calyx sometimes externally pilose, otherwise glabrous:
   2. Leaves with a pair of small immersed marginal glands just above the base, but without macular glands on the lamina; stamens ± 40, filaments pilose at base . . . . **B. marmorata**
   2. Leaves without immersed marginal glands, but with a number of suborbicular macular glands scattered over the lamina; stamens 10–15, glabrous . . . . **B. parviflora**

Baloghia lucida *Endl.*, Prodr. Fl. Norfolk.: 84 (1833) & Iconogr. Pl.: t. 122, 123 (1838); Baill. in Adansonia 6: 206 (1866); Benth.: 148 (1873); Moore & Betche: 79 (1893); Bailey: 1439 (1902); Maiden, For. Fl. N.S.W. 1: 165, t. 28 (1904); Pax & Hoffm. iii: 13 (1911); Anderson: 220 (1968): Baker, Hardw. Austr.: 362 (1919); Francis: 231, figs. 135, 136 (1970); Hyland: 62 (1971); Beadle, Evans & Carolin: 253 (1972). Type: 'Crescit in insula Norfolk, circa finem Januarii florens', *Ferd. Bauer* (W?).

Codiaeum lucidum (Endl.) Muell. Arg.: 1116 (1866); F. Muell., Fragm. 6: 182 (1868).

Q (CK, NK, SK, DD, MO); **NSW** (NC, CC)—Norfolk I., Lord Howe's I., New Caledonia.

Shrub or tree to 15 m high, in rain-forest on red soil, or on ground covered with loose rocks or angular basalt scree rubble, or on conglomerate, up to 300 m.

Readily distinguished from the other two Australian species by the strictly opposite leaves and the frequently white-puberulous tips or margins of the calyx-lobes.

Baloghia marmorata *C. T. White* in Proc. Roy. Soc. Queensl. 53: 226 (1942); Francis: 231 (1970). Type: Q., Tamborine Mt, 1927, *C. T. White* 3588 (BRI).

Q (MO); **NSW** (NC)—Endemic.

Small or medium tree in rain-forest on rich basaltic soils at 540 m.

Glabrous (except for bracts and perianth); the leaves cuneate-oblanceolate or obovate, up to 15 cm long, coriaceous, prominently nerved, with a pair of small but distinct marginal glands just above the base; petiole 0·5–6 cm long. Inflorescence a very abbreviated raceme, borne on an elongate lateral peduncle up to 11·5 cm long, leafless except for a pair of apical opposite elliptic leaves or leaf-like bracts 2·5–5 × 1–2 cm, the flowers subtended by conspicuous ovate or spathulate membranaceous bracts up to 2 × 4 cm, externally glabrous, internally densely sericeous, ciliolate on the margin. Male flowers: pedicel 5 mm long, sepals and petals 5, externally pilose, stamens about 40, filaments long-pilose at the base, disk annular, lobed, fleshy, purplish. Female flower: pedicel 1–1·5 cm long, sepals and petals unknown, ovary glabrous. Capsule 3–4-locular, up to 2 cm diam., seeds 1·5 cm long, mottled irregularly with dark red.

Baloghia parviflora *C. T. White* in Proc. Roy. Soc. Queensl. 53: 227 (1942); Hyland: 73 (1971). Type: Q., Mt Spurgeon, 1936, *C. T. White* 10546 (BRI).

Q (CK)—Endemic.

Tree to 20 m tall, in rain-forest at 700–900 m.

Glabrous or young parts sparsely pilose. Leaves cuneate-oblanceolate or obovate, up to 15 cm long, coriaceous, prominently nerved, bearing on the lower surface a few irregularly scattered elliptic or orbicular ochraceous macular glands 1–2 mm diam.; petiole 1–7·5 cm long. Inflorescences axillary, racemose, borne on slender flattened or angled peduncles; the male 1–7(–12) cm long, occasionally bearing 1 or 2 leafy bracts (up to 4·5 × 2 cm) at the apex, the flower-bearing rhachis 1–3 cm long, angled, brown when dry, rather closely beset with small spreading persistent rigid ovate-triangular bracts 1–2 mm long. Flowers present only in the distal 1–3 bracts; pedicels thick, 1–2 mm long; sepals 3·5 mm long, petals 5 mm long, somewhat fleshy, white; stamens 10–15, glands fleshy. Female inflorescence mostly short, with 1–2 flowers at the apex; ovary glabrous; fruit unknown.

C. T. White (*l.c.*) reported the presence of 2 bracteoles beneath the calyx, but I could find no sign of these, and think they may have been accidental supernumerary sepals.

The species is scattered but locally common on the Atherton Tableland, from the Mossman region to Atherton and Mt Bellenden-Ker.

Breynia *J. R. & G. Forst.*

Scarcely distinct from *Sauropus* Bl., but often blackening in drying; male calyx almost always obconic or turbinate; filaments united in a column, anthers narrow and adnate to the column; styles short and erect, simple or bifid.

1. Leaves obtuse or acute . . . . **B. cernua**
1. Leaves rounded at apex . . . . **B. oblongifolia**

Breynia cernua (*Poir.*) *Muell. Arg.*: 439 (1866); Benth.: 113 (1873); Bailey: 1427 (1902); J. J. Sm.: 180 (1910); Gardner: 72 (1931). Type: 'Cette plante croît dans les Indes', *leg.*? (P).

Phyllanthus cernuus Poir. in Lam., Encycl. Méth., Bot. 5: 298 (1804).
Melanthesa cernua (Poir.) Decne in Nouv. Ann. Mus. (Paris) 3: 483 (1834).
?*Breynia stipitata* Muell. Arg.: 442 (1866); Baill. in Adansonia 6: 344 (1866); Benth.: 114 (1873); Bailey: 1427 (1902); Ewart & Davies: 165 (1917). Type: Q., Sweers Island, *Henne* (G–DC, MEL).
Breynia muelleriana Baill. in Adansonia 6: 344 (1866); Benth.: 114 (1873), *in obs.*; **synon. nov.** Type: Q., Rockingham Bay, 1864, *Dallachy* (G–DC, MEL).
?*Breynia rhynchocarpa* Benth.: 114 (1873); Ewart & Davies: 165 (1917). Type: W.A., King's Sound, *Hughan* (K).
?*Breynia rumpens* J. J. Sm. in Nova Guinea 8: 227 (1910). Type: W New Guinea, Lake Sentani, 1903, *Atasrip* 241 (BO).

WA (K); **NT** (DG); **Q** (CK)—Lesser Sunda Is., New Guinea, Solomon Is., New Hebrides.

Shrub or tree to 8 m tall, in rain-forest, or in open forest on edge of open grassy depression, or on edge of sedge woodland swamp with *Melaleuca*, or on bank of tidal river, or on stable sand-dunes, up to 215 m.

The inclusion of *B. stipitata* and *B. rhynchocarpa* in the above synonymy is entirely provisional, and reflects a policy of despair. I confess to being quite defeated by the baffling variation of the genus *Breynia*. There seems to be little or no correlation between leaf-shape, calyx size and shape, and characters of the fruit. Field studies and possibly cultivation experiments will probably be necessary before the pattern of variation can be properly understood. The genus seems to be in a condition of extreme phenotypic instability.

Breynia oblongifolia (*Muell. Arg.*) *Muell. Arg.*: 440 (1866); Benth.: 114 (1873); Moore & Betche: 75 (1893); Bailey: 1427 (1902); Ewart & Davies: 165 (1917); Beadle, Evans & Carolin: 251 (1972); Airy Shaw in K.B. Add. Ser. VIII: 40 (1980). Syntypes: N.S.W., 'In Nova-Hollandia ad Port Jackson (*d'Urville* in hb. DC.! *Sieb.* hb. DC.! *Sieb.* hb. Flor. Nov. Holl. n. 566!), in insula Timor (hb. DC.!)' (G–DC).

Melanthesa rhamnoides β oblongifolia Muell. Arg. in Linnaea 32: 73 (1863).
B. cinerascens Baill. in Adansonia 6: 344 (1866), *pro nom. nov.* (*illegit.*). Type as above.

var. **oblongifolia**

Q (CK, NK, LT, WB, MO); **NSW**—New Guinea.

Shrub to 3·5 m high, on rocky slopes in rain-forest, on sandstone, granite, basalt, or shale, or in vine thickets or in sandy creek beds, or on grey sandy soil or on a gravelly clay ridge in open *Eucalyptus* forest, up to 780 m.

Not sharply distinct from the *B. cernua* complex; only differing in the narrow or narrower leaf-outline and in the fact that a form with a broad calyx is apparently never developed.

var. **suborbicularis** *Airy Shaw* in K.B. Add. Ser. VIII: 40 (1980). Type: Q., Port Curtis, near Marlborough Homestead, 1963, *Lazarides* 6883 (K).

Q (NK, SK, PC)—New Guinea.

Shrub to 2 m high, on shallow red pebbly soil with lancewood (*Acacia crassicarpa* A. Cunn. ex Benth.).

Leaves suborbicular or very broadly elliptic with a rounded apex. An extreme form, not sharply distinct from var. *oblongifolia*.

Bridelia *Willd.*

Closely related to *Cleistanthus*, differing principally in the venation of the leaves, the main lateral nerves being usually rather close, straight and parallel (very rarely lax and reticulate as in *Cleistanthus*), and in the 1–2- (rarely 3-)locular, drupaceous fruit.

1. Fruit 1-locular, seeds with a deep longitudinal furrow; petals minute; leaves thin, up to 20 × 9 cm . . . . **B. penangiana**
1. Fruit 2-locular, seeds plano-convex; petals larger, imbricate:
  2. Flower-glomerules many-flowered; fruit 4–6 mm diam. (very variable in pubescence, size and shape of leaves, venation, etc.). . **B. tomentosa**
  2. Flower-glomerules 1–3-flowered:
    3. Leaves 5–15 cm long, glabrous, sometimes glaucous beneath, with 10–13 pairs of nerves; fruit 8–10 mm diam.. . . . . **B. exaltata**
    3. Leaves 2–6 cm long, ± pubescent beneath, not glaucous, with 8–10 pairs of nerves; fruit 4–5 mm diam. . . . . . **B. leichhardtii**

Bridelia exaltata *F. Muell.*, Fragm. 3: 32 (1862); Benth.: 119 (1873); Moore & Betche: 76 (1893); Bailey: 1410 (1902); Jabl. viii: 65 (1915); Baker, Hardw. Austr.: 356 (1919); Anderson: 218 (1968); Francis: 226, figs. 133, 134 (1970). Type: N.S.W., Clarence River, *Beckler* (MEL).

B. ovata var. *exaltata* (F. Muell.) Muell. Arg.: 495 (1866).
[*Amanoa ovata* sec. Baill. in Adansonia 6: 336 (1866), *non Bridelia ovata* Decne.]

Q (WB, MO); **NSW** (NC)—Endemic.

Shrub to large tree to 30 m tall, in rain-forest up to 600 m.

Entirely glabrous. Leaves oblong to lanceolate-elliptic, up to 10 × 3·5(–4) cm, rounded or minutely cordate at base, gradually narrowed to an obtuse or subacute apex, chartaceous; margin distinctly undulate, with a marginal nerve; lateral nerves 10–13 pairs. Drupe globose, 8–10 mm diam., glabrous.

Bridelia leichhardtii ('leichhardi') *Baill.* [Ét. Gén. Euphorb.: 584 (1858), *nomen*] *ex Muell. Arg.*: 499 (1866); Jabl. viii: 65 (1915); Francis: 227 (1970). Type: Q., Moreton Bay ['Mt Cameroons'], *Leichhardt* (P).

Amanoa leichhardti (Baill.) Baill. in Adansonia 6: 336 (1866).
A. faginea Baill., *l.c.*: 336 (1866). Syntypes: Q., Rockhampton, Keppel Bay, Frenchmen Creek, 1863, *Dallachy* 17, 259; *Thozet* 76, 192 (MEL).
Bridelia faginea (Baill.) F. Muell. ex Benth.: 120 (1873); Bailey: 1411 (1902).

Q (NK, SK, LT, PC, BT, WB, DD, MO); **NSW**—Endemic.

Tree to 12 m high, in monsoon forest or rain-forest on steep slopes with shallow soils over volcanic rocks up to 820 m.

Branchlets slender, 'twiggy', often nodulose over short sections with crowded persistent bracts or perulae, pubescent when young. Leaves ovate or elliptic, up to 7 × 3 cm, finely pilosulous beneath, the nerves finely but distinctly elevate-reticulate on the upper surface as well as the lower. Drupe globose, 5 mm diam.

Bridelia penangiana *Hook. f.* in Hook. f., Fl. Brit. Ind. 5: 272 (1887); Jabl. viii: 75 (1915); Airy Shaw in K.B. 26: 229 (1971) & 31: 382 (1976) & K.B. Add. Ser. IV: 64 (1975). Type: Malaya, Penang, Government Hill, 1885, *Curtis* 527 (K).

B. minutiflora Hook. f., *l.c.*: 273 (1887) ('*B. micrantha*' in Herb.); Hyland: 71 (1971). Type: Burma, Tenasserim, Mergui, *Griffith* 867 (K).

Q (CK, NK)—SE Asia and throughout Malesia to New Guinea and the Solomon Is.

Tree of 10–15 m, in rain-forest up to 80 m.

Leaves 6–20 × 2·5–9 cm, broadly oblong to elliptic-oblong, cuspidate or shortly acuminate, thin in texture, glabrous or minutely puberulous beneath. Flowers very small, in dense tight axillary glomerules. Fruits shortly oblong or ellipsoid, often with a persistent style.

A form with a definitely pubescent lower leaf-surface, var. *subnuda* (K. Schum. & Lauterb.) Airy Shaw, is known in New Guinea and might well occur in northern Queensland.

Bridelia tomentosa *Bl.*, Bijdr.: 597 (1825); Benth.: 120 (1873); Bailey: 1411 (1902); Jabl. viii: 58 (1915); Gardner: 72 (1931); Airy Shaw in K.B. 26: 231 (1972) & 31: 382 (1976) & K.B. Add. Ser. IV: 65 (1975). Type: Java, 'in montanis Bantam et Buitenzorg', *Blume* (BO).

[*B. monoica* sec. Merr. in Philipp. Journ. Sci. 13, Bot.: 142 (1918) & in Trans. Amer. Philos. Soc. II, 24(2): 234 (1935); Backer & Bakh. f.: 475 (1963); vix *Clutia monoica* Lour.]

var. **tomentosa**

Amanoa tomentosa (Bl.) Baill. in Adansonia 6: 336 (1866).
B. tomentosa var. *ovoidea* Benth.: 120 (1873); Bailey: 1411 (1902); Jabl. viii: 60 (1915). Type: N.T., Wood Island, *Gulliver* (K).

NT (DG)—NE India, SE Asia, S. China, Formosa and throughout Malesia to New Guinea.

Shrub or small tree to 5 m tall, in monsoon forest or mixed scrub on coastal dunes or truncated lateritic podsol, or in low rain-forest scrub in podsolic soil in sandstone plateau depression, at low altitude.

The slender branchlets, rather small, thin, bluntly subacuminate leaves with a marginal nerve, and small globose fruits are characteristic of this species. The thinly pubescent condition of the typical form merges gradually into the completely glabrous state of the following variety.

var. **glabrifolia** (*Merr.*) *Airy Shaw* in K.B. 31: 383 (1976).

B. glabrifolia Merr., Enum. Philipp. Fl. Pl. 2: 422 (1923). Type: Philippines, Manila, *Gaudichaud* (G–C).

[*B. tomentosa* var. *lanceaefolia* sec. Muell. Arg.: 502 (1866), *non B. lanceaefolia* Roxb.]

[*B. ovata* sec. Benth.: 120 (1873); Jabl. viii: 61 (1915); *non* Decne.]

[*B. lancaefolia* sec. Jabl. viii: 60 (1915), *non B. lanceaefolia* Roxb.]

NT (DG); **Q** (CK)—Lesser Sunda Is., Philippines.

Shrub to 4·5 m high, in evergreen forest on sandstone escarpment, or in red sandy soil on gentle slopes with mixed scrub, or forming thickets in savanna forest adjacent to rain-forest, at 20–30 m.

Plant quite glabrous.

var. **trichadenia** *Muell. Arg.*: 501 (1866); Jabl. viii: 60 (1915); Airy Shaw in K.B. 31: 383 (1976). Type: N.T., Timber Creek, Victoria River, *F. Mueller* (G–DC, MEL).

[*B. tomentosa* Bl. sec. Benth.: 120 (1873), quoad specim. *F. Mueller*.]

NT (VR)—? Timor.

Shrub to 1·2 m, on crest of quartzitic hill or in *Eucalyptus tetrodonta* forest on truncated lateritic podsol at low altitude.

Leaves broadly obovate, up to 9 × 5·5 cm, rounded at base and apex; ♀ calyx minutely puberulous; disk pubescent.

var. **eriantha** *Airy Shaw* in Kew Bull. 31: 384 (1976). Type: N.T., N of Pine Creek Township, 1965, *Lazarides & Adams* 145 (K).

NT (DG)—Endemic.

Shrub to 1·2 m on granite hill with *Eucalyptus tetrodonta*, *Sorghum intrans*, and mixed trees and shrubs, or in broadleaf scrub on sandstone conglomerate outlier, at low altitude.

Leaves as in var. *trichadenia*, sometimes subcordate at base; calyx densely pubescent.

Calycopeplus *Planch.*

Shrubs or undershrubs, with terete, angled or flattened stems and branches, the latter opposite, virgate and usually leafless at time of flowering. Leaves opposite or whorled, linear-lanceolate, entire, quickly deciduous;

petiole present or not; stipules minute or obsolete. Flowers in open or campanulate 4-lobed involucres (or cyathia), sometimes with small glands between the lobes. Male flowers arranged in 4 clusters of 4 or more within the cyathium and opposite its lobes, each one subtended and more or less embraced by a bract, the outer 1 or 2 much enlarged and enclosing the cluster; each flower consisting of a single pedicellate achlamydeous stamen, with a well-marked articulation between pedicel and filament; anther with 2 parallel thecae opening longitudinally. Female flower solitary in the centre of the cyathium, pedicellate, with a 4- or 6-lobed perianth (or involucel). Ovary sessile within the perianth, 3-locular, loculi 1-ovulate; styles 3, free or shortly connate, entire or bilobed. Capsule smooth, glabrous; seeds smooth, carunculate.

The genus appears to consist of the three species *C. ephedroides* Planch. (*C. helmsii* F. Muell. & Tate) (WA, NT[?], SA) (stems terete), *C. casuarinoides* L. S. Smith (NT, Q) (stems angled), and *C. marginatus* Benth. (WA) (stems flattened and winged). It is possible, however, that one or two undescribed species remain to be recognized.

Choriceras *Baill.*

Shrubs or small slender trees; branchlets slender, puberulous. Leaves strictly opposite, narrowly ovate to oblong-elliptic, 3–14 cm long, finely crenulate or subentire, coriaceous, glossy and glabrous above, thinly pilosulous beneath, shortly petioled; stipules shortly subulate, very quickly caducous. Flowers monoecious; inflorescences axillary, cymose, ♂ or ♀ or sometimes ♂♀. Male flowers densely fascicled (congested-cymose), borne on rather short, glabrous, filiform pedicels subtended by minute bracts; sepals 2 + 2 or 3 + 3, orbicular, membranous, glabrous; disk 0; stamens 4–6, arising from a fulvous-pilose receptacle, filaments filiform, much exserted, anthers small, globose; pistillode small, fulvous-pilose. Female flowers 1–3 together on a short common peduncle; pedicels stout, almost glabrous, bracts triangular, acute, sericeous; sepals 3 + 3, the outer triangular-ovate, acute, dorsally keeled, the inner minute, narrow; ovary of 3(–4) ovoid, only shortly connate, biovulate carpels, densely sericeous, the styles widely separated, divaricate, recurved, the uncinate apex often lost with age. Capsule profoundly tricoccous, minutely granular, thinly greyish-pilose, crowned by the erect conical bases of the styles; seeds small, ovoid, smooth, ochraceous.

1. Leaves 3–6(–8) × 1–3 cm, closely crenulate-serrate . . **C. tricorne**
1. Leaves 7–14 × 3–7 cm, subentire . . . . **C. majus**

Choriceras majus *Airy Shaw* in Muelleria 4: 220 (1980). Type: Q., 1977, *Hyland* 9365 (K).

Q (CK)—Endemic.

Small tree by creek in rain-forest at 5–450 m.

Distinguished from the common *C. tricorne* by the much larger leaves (up to 14 × 7 cm) and their almost entire margin, and by the rain-forest habitat. The leaves are superficially similar to those of *Whyanbeelia terrae-reginae*, but the midrib is raised on the upper surface, whilst the lower surface is dull and slightly glaucescent. In *Whyanbeelia* the midrib is impressed above and the lower surface is somewhat shining.

The species is so far known only from a small area north of the Daintree River.

Choriceras tricorne (*Benth.*) *Airy Shaw* in K.B. 14: 356 (1961) & 16: 344 (1963). Lectotype (present designation): N.T., Port Essington, 1818 (fruiting), *Cunningham* 265 (K) (original syntype: Q., Rockingham Bay, *Dallachy* (K)).

Dissiliaria tricornis Benth.: 91 (1873); Bailey: 1430 (1902); Ewart & Davies: 167 (1917); Pax & Hoffm. xv: 292 (1922); Specht: 461 (1958).

Choriceras australiana [sic] Baill. in Adansonia 11: 119 (1873) & Hist. Pl. 5: 145 (*in adnot.*), 240 (1874) & Dict. Bot. 10: 28 (1886). Type: N.T., Raffles Bay (E of Port Essington, Coburg Penins.), *Leguillou* (P).

NT (DG); **Q** (CK, ?NK)—Papua.

Shrub or slender tree to 10 m high, very common in sandy soil in heathland, in *Agonis* scrub on sandy ridges, in wallum scrubs by a lagoon, in a ravine in a sandstone escarpment, in *Ptychosperma elegans* forest, in tall *Eucalyptus tetrodonta* open forest on deep yellow earth, in low rain-forest—monsoon forest, or in closed canopy forest along creeks, up to 600 m.

In Queensland this species extends from Cape York to a point south of Cooktown. An early collection of Dallachy (part of Bentham's type material of *Dissiliaria tricornis*) is localized as from 'Rockingham Bay', but the plant seems never again to have been collected from so far south as this, and the locality should probably be regarded as suspect, pending confirmation from modern gatherings.

Claoxylon *Juss.*

Shrubs or small trees, dioecious or rarely monoecious, with alternate, often long-petioled, mostly ± dentate leaves, often scabrous on the upper surface; stipules small or minute; all parts of the plant liable to turn purplish on drying. Inflorescences elongate or abbreviated, simple, rarely branched. Male flowers (1–many in each bract): calyx closed in bud, splitting valvately into 2–4 segments; petals 0; stamens 10–200, free, with almost free, narrow, erect, basifixed anther-thecae, and numerous, narrow, erect juxta-staminal glands interspersed among the filaments; pistillode 0. Female flowers (mostly solitary in each bract): sepals mostly 3, shortly connate; hypogynous disk 2–3-lobed; ovary 2–3(–4)-locular; loculi 1-ovulate; styles short, papillose, spreading. Capsule 2–3(–4)-coccous, cocci mostly deeply separated; seeds globose, with a fleshy outer and a hard foveolate-reticulate inner layer (testa).

1. Monoecious; leaves long and narrow, 10–25 × 1–2 cm, firmly chartaceous, smooth and slightly glossy; petiole 3–10 mm **C. angustifolium**
1. Dioecious; leaves ovate or elliptic, up to 20 × 10 cm; petiole up to 10 cm:

2. Lower leaf-surface rather smooth, the ultimate veins not forming a minute reticulum; leaves always narrow-ovate or elliptic-oblong; capsule always puberulous . . . . **C. australe**
2. Lower leaf-surface dull, the ultimate veins minutely but distinctly reticulate; leaves occasionally broad-ovate; capsule sometimes glabrous **C. tenerifolium**

Claoxylon angustifolium *Muell. Arg.* in Linnaea 34: 165 (1865) & in DC.: 786 (1866); Benth.: 129 (1873); Pax & Hoffm. vii: 125 (1914); Bailey: 1441 (1902). Type: Q., Port Denison, *Fitzalan* (G–DC, MEL).

Mercurialis angustifolia (Muell. Arg.) Baill. in Adansonia 6: 322 (1866).

Q (CK, NK, SK)—Endemic.

Shrub of 2–3 m in dry rain-forest at 80–680 m.

Very distinct in its glabrous, usually strap-shaped or broad-linear leaves, up to 25 cm long and only 1–1·5(–2) cm wide; more rarely elongate-elliptic, up to 14 × 3 cm; texture firmly chartaceous; margin distantly and acutely or obtusely dentate; petiole 3–6(–10) mm long, adpressed-puberulous. Inflorescences very short and slender, 1–2·5 cm long, consisting either of a few male flowers, a few males with one female, or a solitary female flower; the rhachis and male pedicels adpressed-puberulous, but the female pedicels (up to 6 mm long) glabrous. Stamens 9–12; juxta-staminal glands glabrous. Ovary trilocular, minutely puberulous; styles very small, ± conical, depressed. Capsule 4–5 mm diam.

Claoxylon angustifolium is probably related to *C. nervosum* Pax & Hoffm., a scarce endemic of eastern New Guinea.

Claoxylon australe *Baill.* [Ét. Gén. Euphorb.: 493 (1858), *nomen subnudum*] *ex Muell. Arg.*: 788 (1866); Benth.: 130 (1873) (incl. vars. *laxiflora* Benth. & *dentata* Benth., *l.c.*: 131); Moore & Betche: 77 (1893); Bailey: 1442 (1902); Pax & Hoffm. vii: 116 (1914); Anderson: 218 (1968); Francis: 230 (1970); Beadle, Evans & Carolin: 251 (1972). Type: Nouvelle Hollande, *Leichhardt* (P).

Mercurialis australis (Baill.) Baill. in Adansonia 6: 322 (1866).

Q (NK, PC, WB, DD, MO); **NSW** (NC, CC, SC)—Endemic.

Tree to 9 m high, on humus in rain-forest up to 840 m.

Leaves obovate-oblong, up to 13 × 5 cm, obtusely or more rarely acutely sinuate-dentate or sometimes subentire, firmly chartaceous, variously puberulous when young, practically glabrous when mature, cuneate or rarely narrowly rounded at base, rounded or sometimes acute at the apex; petiole 1–4 cm long. Male inflorescence up to 8 cm long, female up to 4 cm. Stamens 15–25; juxta-staminal glands pilose at apex. Ovary puberulous; styles short, subulate, spreading. Capsule tricoccous, 5–6 mm diam., minutely puberulous.

Claoxylon tenerifolium (*Baill.*) *F. Muell.*, Fragm. 6: 183 (1868); Benth.: 130 (1873); Bailey: 1441 (1902); Pax & Hoffm. vii: 114 (1914); Airy Shaw in K.B. 31: 390 (1976). Syntypes: Q., Rockhampton, Thozet Creek, 1863, *Dallachy* 137 (P, MEL); Broad Sound, *Bowman* 151 (P, MEL).

Mercurialis tenerifolia Baill. in Adansonia 6: 323 (1866),
Claoxylon hillii Benth.: 131 (1873); Bailey: 1442 (1902); Pax & Hoffm. vii: 114 (1914); Hyland: 52 (1971). Lectotype (Airy Shaw 1976): Q., Albany Island, Cape York, *W. Hill* 4 (K) (other original syntypes: Q., Cape York, *McGillivray*, *Daemel* (K); Rockingham Bay, *Dallachy* (K)).
C. australe var. *latifolia* Benth.: 131 (1873); Bailey: 1442 (1902); Pax & Hoffm. vii: 116 (1914). Type: Q., Rockhampton, *O'Shanesy* (K).
C. indicum var. *novoguineense* J. J. Sm. apud Valet. in Bull. Dép. Agric. Ind. Néerl. 10: 26 (1907) & in Fedde, Rep. 5: 382 (1908) & in Nova Guinea 8: 231 (1910); Pax & Hoffm. vii: 110 (1914). Type: W New Guinea, Merauke, 1904, *Koch* 25 (BO).
C. delicatum Airy Shaw in K.B. 20: 32 (1966). Type: Papua, Peria Creek, 1953, *Brass* 24082 (K).

NT (DG); **Q** (CK, NK, PC)—New Guinea.

Shrub or small tree to 15 m tall, in sink hole in sandstone rain-forest (NT), or on sand cay, or in gallery rain-forest or deciduous vine thicket, up to 860 m. Fig. 1D.

Distinguished by its thin but somewhat sappy leaves, in which the minor nerves form an exceedingly fine, ± immersed reticulum, especially on the under side. The whole plant tends to assume a purplish colour on drying. The capsules may be either glabrous (*tenerifolium* s. str.) or puberulous (*C. hillii*). The species is rather frequent in New Guinea; it is closely related to the common W Malesian *C. indicum* Hassk.

Cleidion *Bl.*

Shrubs or small trees, glabrous or with simple pubescence. Leaves alternate, often obovate and distinctly crenate, or oblanceolate and very shortly petioled; stipules very inconspicuous. Flowers mostly dioecious. Male inflorescence glomerate-spicate, sometimes greatly elongate; calyx closed in bud, valvately 3–4-partite at anthesis; disk 0; stamens very numerous, the anthers arranged in regular vertical series, 4-locellate, connective shortly produced; pistillode 0. Female inflorescence either elongate fasciculate-racemose, or consisting of a solitary, axillary, long-pedicelled flower; sepals 3–5, imbricate; disk 0; ovary 2–3-locular; styles conspicuous, elongate-filiform, deeply bifid, often connate below. Capsule 2–3-locular, when solitary and axillary borne on a rigid flattened pedicel; seeds subglobose, testa often mottled.

Cleidion spiciflorum (*Burm. f.*) *Merr.*, Interpr. Rumph. Herb. Amboin.: 322 (1917), *in obs.*, & Enum.: 439 (1923); Airy Shaw in K.B. 26: 234 (1972). Type: India orientalis, *leg.* ?

Acalypha spiciflora Burm. f., Fl. Ind.: 203 (sphalm. '303'), t. 61, fig. 2 (1768).
Cleidion javanicum Bl., Bijdr.: 613 (1825); Baill., Ét. Gén. Euphorb., Atlas: t. 9, figs. 3–5 (1858); Muell. Arg.: 987 (1866); Pax & Hoffm. vii: 290 (1914); Backer & Bakh. f.: 487 (1963); Airy Shaw in K.B. Add. Ser. IV: 74 (1975). Type: Java, 'in monte Bonkok Provinciae Tjanjor', *leg.*? (BO).
Claoxylon spiciflorum (Burm. f.) Baill., Ét. Gén. Euphorb., Atlas: 37, t. 20, figs. 20–24 (1958).

Q (CK)—India, SE Asia and S China, and throughout Malesia to the Bismarcks and Solomon Is.

Small tree to 7 m high, in dry or gallery rain-forest at 60–480 m.

Leaves obovate, up to 20 × 8 cm, cuneate to rounded at base, obtusely cuspidate and mucronate at apex, margin distantly crenate-dentate or subentire, chartaceous, glabrous; petiole slender, up to 5 cm long, glabrous, mostly narrowly pulvinate at apex. Male inflorescence interruptedly glomerate-racemose, up to 20 cm long, rhachis and pedicels minutely puberulous. Sepals 3, stamens 40–70, anthers 4-locellate, connective distinctly or minutely produced. Female inflorescence 1-flowered, solitary, axillary, peduncle c. 2 cm long, elongating to 8 cm in fruit. Sepals 4–5; ovary 2(–3)-locular, sparsely minutely puberulous; styles 2(–3), up to 15 mm long, connate for 2–4 mm, deeply bifid, papillose. Capsule didymous, 2·5 × 1·5 cm, granular, glabrous, peduncle flattened and expanded upwards.

Occasional in the Cape York Peninsula and in the Atherton Tableland.

Cleistanthus *Hook. f. ex Planch.*

Trees or shrubs. Leaves alternate, entire, chartaceous or coriaceous, penninerved (nerves mostly lax and ± reticulate, occasionally parallel as in *Bridelia*), shortly petioled; stipules small or minute, deltoid or setaceous. Inflorescences axillary or extra-axillary, fasciculate-glomerate, sometimes borne on leafless or small-leaved branchlets, monoecious; fascicles ♂ or ♀ or ♂♀. Male flower: calyx tube shortly obconic, segments strictly valvate; petals 5, small or minute, free, unguiculate-spathulate, delicate, often erose at apex; disk annular or shortly cupular, sometimes apparently double or with a double margin; stamens 5, filaments connate below in a column or androgynophore, free and spreading above; anthers ± versatile; pistillode very small, trifid or trilobed, borne at apex of column. Female flower: calyx and petals as male; disk annular or cupular or urceolate; ovary 3-locular, pubescent or glabrous, loculi biovulate, styles 3, ± connate below, simple or bifid or twice bifid. Capsule 3-locular, sessile or pedicelled, sometimes also stipitate within the calyx; seeds trigonous, cotyledons either thin and flat (sometimes folded) or thick and fleshy, endosperm correspondingly copious or sparse.

1. Ovary and fruit glabrous:
 2. Leaves densely minutely golden or coppery-sericeous beneath; capsule stipitate; embryo thick; endosperm almost lacking . **C. myrianthus**
 2. Leaves glabrous beneath; capsule various; endosperm evident:
  3. Leaves 1–4 cm broad, elliptic to narrowly oblong-elliptic, dull, finely reticulate; capsule markedly (2 mm) stipitate . . **C. xerophilus**

3. Leaves 2–8 cm broad, broadly to narrowly ovate, shining above, laxly or less finely reticulate:
 4. Leaves mostly larger; bracts minute; indumentum of bracts golden; capsule shortly (1 mm) stipitate . . . . **C. hylandii**
 4. Leaves smaller; bracts conspicuous; indumentum of bracts chocolate-brown; capsule almost sessile . . . . **C. dallachyanus**

1. Ovary (and often fruit) pubescent; embryo thin:
 5. Capsule stipitate; inflorescence various:
 6. Branchlets slender; leaves small, chartaceous, glabrous or finely whitish-puberulous beneath, with spreading nerves and reticulate venation; capsule 7–8 mm diam. . . . . . **C. cunninghamii**
 6. Branchlets robust; leaves up to 18 cm long, coriaceous, densely coppery-sericeous beneath, with few strongly ascending nerves and relatively inconspicuous venation; capsule 12 mm diam. . . . **C. discolor**
 5. Capsule not stipitate; inflorescences mostly on special leafless or small-leaved branches:
 7. Branchlets and calyces glabrous; bracts often shortly whitish-lanate **C. apodus**
 7. Branchlets, calyces and bracts conspicuously pubescent:
 8. Indumentum ferrugineous; ♀ flower 3–4 mm long **C. semi-opacus**
 8. Indumentum pale or cinnamomeous; ♀ flower 4–5 mm long **C. peninsularis**

Cleistanthus apodus *Benth.*: 122 (1873); Bailey: 1412 (1902); Jabl. viii: 15 (1915). Lectotype (present designation): Q., Cape York, *McGillivray* 417 (K) (syntypes: Cape York, *Daemel* (K); Rockingham ('Rockhampton') Bay, *Dallachy* (K)).

Q (CK, NK)—Papua.

Small tree with drooping branches, to 10 m high, in gallery rain-forest up to 80 m.

A typical species of sect. *Leiopyxis*, with glabrous branchlets, very smooth glabrous leaves, usually greyish-green or greyish-brown when dry, and inflorescences borne on either leafless or leafy branchlets. The glomerules are few-flowered, the calyces glabrous, and the floral bracts white-pilose. Capsule sessile and estipitate, strongly tricoccous, thinly pilose, about 8 mm diam.

Occurs rather frequently along creek-sides from the region of Tully and Rockingham Bay northwards to Cape York, and again in southern Papua.

Cleistanthus cunninghamii (*Muell. Arg.*) *Muell. Arg.*: 506 (1866); Benth.: 122 (1873); Moore & Betche: 76 (1893); Bailey: 1412 (1902); Jabl. viii: 34 (1915); Ewart & Davies: 167 (1917); Anderson: 217 (1968). Syntypes: N.S.W., 'In New South Wales Novae Hollandiae', *Cunningham* 31 & 120 (G–DC).

Lebidiera cunninghamii Muell. Arg. in Linnaea 32: 80 (1863).
Amanoa cunninghamii (Muell. Arg.) Baill. in Adansonia 6: 335 (1866).

?NT (VR); **Q** (SK, PC, WB, MO); **NSW** (NC)—Endemic.

Shrub (sometimes scrambling) or small tree to 7 m high, in rain-forest or dry rain-forest or fringing forest on shallow stony soil up to 800 m alt.

Readily recognizable from its small, thin, veiny leaves, up to 8 × 3·5 cm, with very widely, sometimes almost horizontally spreading main nerves. The shape varies from ovate to lanceolate, the margin is often markedly reflexed, and the lower surface bears a characteristic minute puberulence of short, pale, adpressed hairs pointing in different directions, finally ± glabrescent. The main nerves are often slightly bullately impressed. Flower-glomerules rather few and few-flowered, mostly extra-axillary; calyx glabrous; ovary long-pubescent. Capsule 6–7 mm diam., deeply tricoccous, ochraceous- or ferrugineous-pilose, borne on a slender stipes 2–3 mm long.

The species extends north in Queensland about as far as the latitude of Mackay.

The 'imperfect specimen' from the Northern Territory cited by Bentham ('Frutex arborescens. Ad flumen Victoriae. Oct. 55. Ferd. Mueller') appears certainly to be *C. cunninghamii*, although the capsule is estipitate, but the locality must be regarded as suspect in the absence of later confirmation. It could well be the result of mislabelling.

Cleistanthus dallachyanus (*Baill.*) *Baill. ex Benth.*: 122 (1873); Bailey: 1412 (1902); Jabl. viii: 36 (1915). Syntypes: Q., Rockhampton, *Dallachy* 17; Mt Mueller, *Dallachy*; Port Denison, *Dallachy*; Rockhampton, *Thozet* 337 (all P, MEL).

Amanoa dallachyana Baill. in Adansonia 6: 335 (1866).

Q (CK, NK, ?SK, PC)—Endemic.

Shrub or tree to 6 m high, in monsoon forest or in gallery rain-forest on dark grey loam at 15–20 m.

The shining upper surface of the leaves with their conspicuously raised reticulate venation, combined with the deep chocolate-brown colour of the short, mostly leafless, dense-flowered inflorescences, makes this an easily recognized species. The leaves are mostly broadly ovate and abruptly caudate or gradually narrowed to the apex. In sect. *Australes* (with glabrous ovary) *C. dallachyanus* comes nearest to *C. hylandii*, from which it differs in the much larger, conspicuous floral bracts, the deep brown colour of the whole inflorescence and the ferrugineous rather than golden colour of the pubescence. The capsule is practically estipitate.

The species is known from a few scattered localities from Rockhampton as far north as Bathurst Bay.

Cleistanthus discolor *Summerh.* in Bull. Misc. Inf., Kew, 1928: 144 (1928); Hyland: 38 (1971). Type: Q., Kuranda, 1927, *Du Rietz* 7275 (K).

Q (CK)—Endemic.

Small tree in rain-forest on metamorphic rocks at 360 m.

A species exhibiting extreme heterophylly as between the floral leaves subtending the flower-glomerules, which may be narrowly elliptic and only 15 × 4 mm, and the mature leaves of the main branches, which may be broadly elliptic and up to 18 × 8·5 cm in size. The leaves are stiffly coriaceous, glabrous above and densely adpressedly grey or rufescent-pubescent beneath; nerves 4–5 pairs only, steeply ascending; petiole thick, 5–6 mm long. Floral bracts ferrugineous-pubescent. Calyx glabrous. Capsule 12–13 mm diam., rather shallowly tricoccous, glabrous, 2–3 mm stipitate. The flowering branchlets are often flattened or irregularly angled.

Cleistanthus discolor is a scarce endemic, known only from a very few collections from between Cairns and Mareeba. On technical characters it falls, with *C. cunninghamii*, into sect. *Chartacei*, but the two species cannot be said to be closely related.

Cleistanthus hylandii *Airy Shaw* in K.B. 31: 379, 381, *in clavi* (1976). Type: Q., Claudie River, 1974, *Hyland* 3099 RFK (holotype K).

Q (CK)—Endemic.

Small tree to 8 m high, in gallery rain-forest or dry scrub up to 80 m.

Leaves ovate, very variable in size, up to 16 × 8 cm, glabrous beneath, green when dry. The species differs from the (perhaps only distantly) related *C. dallachyanus* (Baill.) Benth. in the less reticulately raised nervation of the upper leaf-surface, the golden-rusty rather than fuscous-rufous pubescence of the inflorescences, the very small and inconspicuous bracts, the shortly dichotomously branched apices of the styles, and the more distinctly (1 mm) stipitate capsule. The inflorescences are mostly borne on leafless branchlets but are sometimes also axillary. The ovary is quite glabrous and the cotyledons are broad and thin as in the remainder of the sect. *Australes*.

Cleistanthus hylandii is so far known only from the region of Iron Range, Scrubby Creek and Rocky River.

Cleistanthus myrianthus (*Hassk.*) *Kurz*, For. Fl. Brit. Burma 2: 370 (1877); Airy Shaw in K.B. 26: 237 (1971) & 31: 378, 381, *in clavi* (1976) & K.B. Add. Ser. IV: 83 (1975). Type: W Java, Bantam, S coast, 1841, *Hasskarl* (BO). [Added in proof: Jabl. viii: 37 (1915)].

Nanopetalum myrianthum Hassk. in Verhand. Kon. Akad. Wetensch. Amsterd. 24: 140 (1855) & in Bot. Zeit. 14: 803 (1856) & in Flora 40: 534 (1857); Muell. Arg.: 510 (1866).

Q (CK)—SE Asia and throughout Malesia to the Solomon Is.

Small tree to 7 m tall, in rain-forest up to 390 m.

Readily recognized among the Australian species by the very fine, adpressed, shining coppery pubescence of the lower leaf-surface. The leaves are coriaceous, mostly ovate-lanceolate, but sometimes elliptic or obovate, up to 15 × 5·5 cm. Inflorescence either leafy or leafless; ovary glabrous; capsule stipitate (but scarcely pedicellate); embryo thick, with very scanty endosperm (one of the chief distinguishing characters of Sect. *Nanopetalum*, to which *C. myrianthus* belongs).

So far known only from the region of Gap Creek, Noah Creek and Alexandra, north of Cooktown.

Cleistanthus peninsularis *Airy Shaw & Hyland* in Muelleria 4: 222 (1980). Type: Q., near Musgrave, 1973, *Hyland* 6927 (K).

C. semiopacus var. *curvaminis* Airy Shaw in K.B. 31: 380 (1976). Type: Q., The Bend, near Coen, 1962, *L. S. Smith* 11973 (BRI).

Q (CK)—Endemic.

Shrub or small tree to 10 m tall, in dry monsoon forest or gallery forest at 60–210 m.

Closely related to *C. semiopacus*, but differing in the much less glossy upper surface of the leaves, the paler, cinnamomeous rather than ferrugineous indumentum, and the considerably larger flowers. The species has the densely pubescent ovary of sect. *Leiopyxis*, but the inflorescences are borne on leafy branches, unlike most species of that section. The capsule and seeds are so far unknown.

Cleistanthus peninsularis occurs from near Musgrave, at the southern end of the Cape York Peninsula (lat. 14° 40′ S), to Cape York itself and the islands of the Torres Strait, and across to Dauan Island, off the coast of Papua, where it is sometimes a dominant tree in the low forest.

Cleistanthus semiopacus *F. Muell. ex Benth.*: 123 (1873); Jabl. viii: 16 (1915); Hyland: 81 (1971); Airy Shaw in K.B. 31: 382, *in clavi* (1976). Type: Q., Rockingham Bay, *Dallachy* (K).

C. semiopacus var. *semiopacus*; Airy Shaw in K.B. 31: 380 (1976).

Q (CK, NK)—Endemic.

Tree to 15 m tall, in rain-forest or dry rain-forest up to 480 m.

Leaves elliptic or narrowly ovate, shortly acuminate, chartaceous or thinly coriaceous (Bentham's 'thickly coriaceous' is misleading), very smooth and glossy above, dull and slightly glaucous or pinkish-pruinose beneath, adpressed coppery-pubescent beneath when young, finally glabrescent. Inflorescences on leafy or leafless branchlets, strongly ferrugineous-pubescent, the glomerules many-flowered. Ovary densely pubescent. Capsule small, strongly tricoccous, 5–6 mm diam., sessile and estipitate, adpressed-rufous-pubescent.

Only known from the Atherton Tableland and the region of Rockingham Bay.

Cleistanthus xerophilus *Domin*: 879 [325] (1927); Airy Shaw in K.B. 31: 381 (1976). Type: Q., between Chillagoe and Walsh River, 1910, *Domin s.n.* (PR).

C. densiflorus C. T. White in Proc. Roy. Soc. Queensl. 55: 82 (1944). Type: Q., Bloomfield River, 1902, *Poland s.n.* (BRI).

Q (CK)—Endemic.

Small shrubby tree to 5 m tall, growing on the edge and higher parts of the bed of creeks in open eucalypt forest at 230–240 m. Fig. 1C.

The semi-rheophytic habit, narrowly elliptic or almost oblong leaves, deep rufous-brown indumentum of young parts and bud-scales, glabrous stipitate ovary and long-stipitate capsule (stipes up to 1 cm long!) are unmistakable characters of this species. The leaves are obtuse, mostly dull and finely reticulate on the upper surface, and only 1–2 cm broad, but occasionally they exceed 4 cm in breadth. The glabrous ovary and thin flat cotyledons of the embryo associate this species with *C. dallachyanus* and *C. hylandii* in sect. *Australes*.

Cleistanthus xerophilus has been obtained in a number of localities from the McIlwraith Range to the Chillagoe region.

Codiaeum *Bl.*

Shrubs or small trees, without stellate hairs. Leaves alternate, entire, mostly oblong or obovate, but the lamina sometimes displaying curious modifications, twisting, reduction of surface, etc., coriaceous to membranaceous, penninerved, petiolate; stipules minute or obsolete. Inflorescences laxly elongate-racemose, usually borne in the axil of a small, orbicular, modified leaf, usually unisexual and monoecious, male flowers ± fascicled on filiform pedicels, females solitary on a shorter stouter pedicel in the axil of each bract. Male flower: sepals (3–)5(–6), free, imbricate; petals small or minute, ± membranous, occasionally 0; disk-glands small, free, alternating with petals; stamens 15–100, free; pistillode 0. Female flower: sepals as male, but smaller; disk entire, shortly cupular; ovary trilocular, glabrous or adpressed-pubescent; styles mostly elongate, simple (rarely bipartite), terete, recurved. Capsule 3-coccous, smooth; seeds shining, marbled.

1. Leaves coriaceous; ovary glabrous . . . . **C. variegatum**
1. Leaves membranaceous [ovary pubescent?] . **C. membranaceum**

Codiaeum membranaceum *S. Moore* in Journ. Linn. Soc., Bot. 45: 219 (1920); Airy Shaw in Muelleria 4: 237 (1980), *q.v.* for discussion. Type: Q., Cape York, 1868, *Daemel* (BM).

Q (CK)—Endemic?

Tree (bole 10 cm d.b.h.), in rain-forest at 30 m.

The status of some of the taxa that have been described in *Codiaeum* needs elucidation from more abundant material. *C. membranaceum*, described from Cape York, differs most obviously from *C. variegatum* in the membranaceous texture of the leaves. If a recent collection from somewhat further south in the Cape York Peninsula (*Hyland* 2919) is correctly referred here, the species is further distinguished by having a pubescent ovary. The validity of these distinctions (and their application also to other species that have been proposed from parts of Malesia) must be tested from further material.

Codiaeum variegatum (*L.*) *Bl.*, Bijdr.: 606 (1825); Baill., Ét. Gén. Euphorb., Atlas: t. 16, figs. 26–35 (1858); J. J. Sm.: 22 (1910). Type: 'Habitat in Amboina' [Rumph., Herb. Amboin. 4: 65, t. 25 (1743)].

Croton variegatus L., Sp. Pl. ed. 3: 1424 (1764).

Codiaeum chrysosticton [Rumph. ex] Spreng., Syst. Veg. 3: 866 (1826); F. Muell., Fragm. 6: 182 (1868); *nom. illegit. superfl.* Type: as for *Croton variegatus* L.

var. **moluccanum** (*Decne*) *Muell. Arg.*: 1119 (1866); Benth.: 147 (1873); Bailey: 1438 (1902); Pax & Hoffm. iii: 24 (1911); Airy Shaw in Muelleria 4:237 (1980).

C. moluccanum Decne, Descr. Herb. Île Timor, in Nouv. Ann. Mus. Paris 3: 485 (1834). Type: Timor, *Riedlé & Guichenot* (P).

Codiaeum obovatum Zoll. in Flora 30: 663 (1847); Baill. in Adansonia 6: 303 (1866). Type: Java (SE), in sylvis litoralibus, *Zollinger* II. 583, Herb. 2435 (P).

Junghuhnia glabra Miq., Fl. Ind. Bat. 1(2): 412 (1859). Type: Java, coast near Kondong Tebo, *Zollinger* 2435 (P).

Croton mirus Domin: 882 [328], t. 31, figs. 1–10 (1927). Type: Q., Harvey's Creek, Jan. 1910, *Domin s.n.* (PR).

Q (CK, NK)—E Malesia, New Guinea, Bismarcks, Pacific Is.

Shrub of 2–3 m, plentiful in rain-forest undergrowth (gallery or floodplain) up to 600 m.

The many variations of this plant are commonly known in tropical gardens under the general term 'croton', but the genus *Codiaeum* is not remotely related to the genus *Croton* L.

Croton *L.*

Trees or shrubs, occasionally rheophytic (one introduced species an annual herb), densely or sparsely clothed with stellate hairs or shining scales, occasionally subglabrous. Leaves alternate or pseudo-verticillate, petiolate, subentire or crenate or dentate or occasionally lobed, penninerved or sometimes palminerved at the base, biglandular at junction of petiole and lamina; stipules minute or shortly filiform, sometimes obsolete. Flowers mostly monoecious. Inflorescences terminal, racemose, androgynaecious, the female flowers sometimes reduced to 1 basal long-pedicelled flower. Male flower: sepals mostly 5, free, imbricate or valvate; petals 5, free, often lanate at the apex; disk-glands small, opposite the sepals; stamens 5–30, mostly lanate at the base, inflexed at the apex in bud; pistillode 0. Female flower: sepals much as in male; petals mostly smaller or vestigial; ovary 3-locular; styles variously divided into 2 or 4 linear or thickened branches or occasionally shortly flabellate. Capsule tricoccous, smooth or shortly muricate; seeds ovoid or ellipsoid, smooth, occasionally sparsely stellate-lepidote.

[Specific distinctions in *Croton* are largely relative and often elusive. It is therefore risky to name specimens without comparing authentically named material.]

1. Mature leaves glabrous or almost so:
  2. Leaves thinly membranaceous-herbaceous:
    3. Leaves with a conspicuously reflexed or revolute margin, shallowly repand-denticulate, often markedly acuminate . . . **C. dockrillii**
    3. Leaves usually with a flat margin, crenate to closely serrate-crenulate, not or scarcely acuminate:
      4. Pubescence thinly and minutely stellate; apical petiolar glands almost sessile, usually close to the base of the lamina . . **C. armstrongii**

4. Pubescence virtually absent; petiolar glands often long-stipitate, relatively distant from lamina . . . . **C. byrnesii**

2. Leaves chartaceous to coriaceous:

5. Leaves coriaceous, glossy, with venation conspicuously raised on the upper surface, inconspicuous below, shallowly crenate-serrate to subentire; petiole 5–12 mm long; ♀ flowers crowded at base of inflorescence; young stems and petioles ochraceous-lepidote with subentire scales . . . . **C. acronychioides**

5. Leaves chartaceous, less shining, venation not conspicuously raised on upper surface; petiole 1–2(–3·5) cm long; ♀ flowers not crowded at base of inflorescence:

6. Leaves usually distinctly and closely but shallowly serrulate, 1–4 cm broad, narrower in general outline:

7. Young branchlets, inflorescence-axes, petioles and leaf-undersurface (when dry) purple or purplish; stellate hairs very sparse **C. verreauxii**

7. Young branchlets, etc., ochraceous-brown when dry; stellate indumentum evident . . . . **C. prunifolius**

6. Leaves subentire or obscurely crenate, 2–6·5 cm broad and broader in general outline:

8. Petioles 2–5 mm long . . . . **C. brachypus**

8. Petioles over 5 mm long:

9. Petioles 5–20 mm long . . . . **C. triacros**

9. Petioles 15–35 mm long . . . . **C.** cf. **storckii**

1. Mature leaves manifestly stellate-pubescent or lepidote beneath:

10. Leaves densely shining silvery- or coppery- or greenish-lepidote beneath:

11. Apical petiolar glands conspicuously stipitate; lepidote scales greenish; leaves up to 16 × 6 cm . . . . **C. capitis-york**

11. Apical petiolar glands not stipitate; lepidote scales pinkish or silvery:

12. Female calyx about 6 mm long; leaves mostly broad-ovate with rounded base . . . . **C. argyratus**

12. Female calyx 2 mm long; leaves mostly narrow-ovate with narrowed base . . . . **C. insularis**

10. Leaves densely or thinly stellate-pubescent, not or scarcely shining beneath:

13. Leaves ovate to orbicular-ovate, often cordate at base:

14. Leaves variously crenate-dentate, strongly 3–7-nerved at the base; stamens 20–20; capsule densely rufous-stellate-tomentose **C. arnhemicus**

14. Leaves subentire or finely crenulate, not or much less strongly 3–5-nerved at base; stamens (so far as known) 10–12:

15. Nerves distinctly incised above; ovary (and capsule?) muricate **C. stockeri**

15. Nerves not incised above; ovary and capsule smooth **C. tomentellus**

13. Leaves linear or narrow-oblong or oblong-elliptic to lanceolate, occasionally obovate, rarely subcordate:

16. Pubescence coarse, spreading, ± ochraceous; styles bifid; leaves obovate or elliptic . . . . **C. densivestitus**

16. Pubescence fine, close, whitish:

17. Leaves 0·3–1·5 cm wide, narrow-oblong, entire or subentire; styles 4–6-fid . . . . **C. phebalioides**
17. Leaves 1·5–5 cm wide, lanceolate- or elliptic-oblong:
    18. Leaves often shortly repand-denticulate, without stipitate glands in the marginal sinuses, blackish above when dry; nerves narrowly incised above; styles 4–6-fid . . . . **C. stigmatosus**
    18. Leaves slightly indentate-crenate, with a small stipitate gland in many of the sinuses, muddy brown when dry; nerves not incised; styles deeply bifid . . . . **C. magneticus**

Croton acronychioides *F. Muell.*, Fragm. 4: 142 (1864); Baill. in Adansonia 6: 300 (1866), *cum descr. amplif.*; Benth.: 127 (1873); Bailey: 1437 (1902); Anderson: 218, 377 (1968); Francis: 227 (1970); Airy Shaw in Muelleria 4: 223 (1980). Syntypes: Q., Fitzroy River, near Rockhampton, *Thozet* (MEL); Broad Sound, *Bowman* (MEL).

C. affinis Maiden & R. T. Baker in Proc. Linn. Soc. N.S.W. II, 9: 160, t. 12 (1894); Francis, *l.c.*: 230 (1970); **synon. nov.** Type: N.S.W., near Tintenbar, Richmond River, *Bäuerlen s.n.* (NSW).

[*C. verreauxii* sec. Airy Shaw in K.B. 31: 386 (1976), quoad synon. tantum, *non* Baill.]

Q (PC); **NSW** (NC)—Endemic.

Tree in rain-forest at 800 m.

Recognizable from its stiffly coriaceous, glossy leaves, with the venation conspicuously raised on the upper surface; the margin may be shallowly crenate-serrate or subentire. The young stems and petioles are ochraceous-lepidote, with subentire scales, and the female flowers tend to be tightly bunched at the base of the inflorescences.

So far known only from the Rockhampton region and the north-east corner of New South Wales.

Croton argyratus *Bl.*, Bijdr.: 602 (1825); Muell. Arg.: 526 (1866); J. J. Sm.: 336 (1910); Backer & Bakh. f.: 476 (1963); Airy Shaw in K.B. 26: 243 (1972) & 31: 385 (1976) & K.B. Add. Ser. IV: 90 (1975). Type: Java, 'in sylvis montium calcareorum Provinciarum occidentalium Javae', *Blume* (BO).

C. schultzii Benth.: 124 (1873). Type: N.T., Port Darwin, 1870, *Schultz* 609 (K).

NT (DG)—SE Asia and W Malesia to Moluccas and Bali.

Shrub of 1 m in monsoon forest or deciduous vine thicket at very low altitude.

When in flower or fruit the relatively large, ± oblong sepals of the female flower (5–6 mm long) distinguish *C. argyratus* clearly from the other lepidote Australian species, *C. insularis* and *C. capitis-york*. From *C. insularis* it is further distinguished by its broader leaves, generally rounded or subcordate at the base, by its shortly pedicelled flowers, and by its slender elongate style-branches. From *C. capitis-york* it differs in the pinkish rather than greenish colour of the lepidote undersurface of the leaves when dry, and in the minute or almost obsolete glands at the base of the midrib (lower surface). There is probably a difference also in the style-branches; see under *C. capitis-york.*

Croton armstrongii *S. Moore* in Journ. Linn. Soc., Bot. 45: 219 (1920); Airy Shaw in Muelleria 4: 224 (1980). Type: N.T., Port Essington, *Armstrong s.n.* (BM).

C. habrophyllus Airy Shaw in K.B. 31: 386 (1976). Type: N.T., Port Darwin, 1870, *Schultz* 680 (K).

[*C. verreauxii* sec. Benth.: 126 (1873), quoad specim. *Brown, Armstrong & Schultz*; Chippendale in Proc. Linn. Soc. N.S.W. 96: 244 (1972); *non* Baill.]

NT (DG)—Endemic.

Shrub of 2–3 m, in monsoon rain-forest on gravelly lateritic soil, or in semi-evergreen vine forest, or on stable coastal dunes, at low altitude.

Distinguished from all other Australian species, except *C. byrnesii*, by its thinly membranaceous leaves, and from *C. byrnesii* by its thin minutely stellate pubescence, by the almost sessile petiolar glands, and by the shorter distance separating these from the base of the lamina.

A critical micro-species from the region of Darwin and the Coburg Peninsula.

Croton arnhemicus *Muell. Arg.* in Linnaea 34: 112 (1865) & in DC.: 599 (1866); Baill. in Adansonia 6: 300 (1866); Benth.: 127 (1873); Bailey: 1437 (1902). Syntypes: N.T., 'in Arnhemsland Novae-Hollandiae septentrionalis', *F. Mueller* (G–DC, MEL); Sea Range, towards the Fitzmaurice, *F. Mueller* (G–DC, MEL); Q., Cape York, *MacGillivray* 514 (K).

C. arnhemicus var. *urenifolius* Baill., *l.c.*: 300 (1866); Benth.: 128 (1873); Bailey: 1438 (1902). Syntypes: Q., Port Denison, Edgecombe Bay, *Fitzalan* (P, MEL); Edgecombe Height, 1863, *Dallachy* (P, MEL).

NT (DG, VR, BT); **Q** (BK, CK, NK)—Endemic.

Shrub or tree to 5 m tall, in open eucalypt forest on gentle slopes in red sandy soil or sandy loam or Tippera clay soil, or grey, gravelly, sandy soil ('florina') on creek flat, or in littoral forest, or in dry rain-forest–monsoon forest, or in deciduous vine thicket on limestone outcrop, up to 480 m.

The dense stellate (but not lepidote) tomentum and the strongly 3–7-nerved leaves with a variously crenate-dentate margin distinguish *C. arnhemicus* from *C. tomentellus*, the only species with which it might be confused. Moreover according to F. Mueller the latter has only about 10 stamens in the male flower, as against 20–30 in *C. arnhemicus*, but I have not yet been able to dissect male material of *tomentellus*.

Croton arnhemicus is scattered over a wide area of northern Australia, and is somewhat variable. I am at present unable to uphold var. *urenifolius* Baill. as a distinct entity, or to propose other formal varieties, but it is not impossible that more abundant material may ultimately warrant this.

Croton brachypus *Airy Shaw* in Muelleria 4: 224 (1980). Type: Q., Tozer Range, 1948, *Brass* 19462 (K).

[*C. mirus* sec. Airy Shaw in K.B. 31: 385 (1976), quoad *Brass* 19462 tantum, *non* Domin.]

Q (CK)—Endemic.

Tree or shrub of 1·5–4 m, in dry rain-forest at 200 m, or in semi-deciduous mesophyll vine-forest along streams, on alluvial soils derived from a mixture of granite and metamorphic rocks, at 200 m, or in heath scrub at 360 m, or characteristic of rain forest undergrowth at 425 m.

Closely related to *C. triacros*, of the Atherton Tableland and Rockingham Bay region, but differing in its conspicuously and consistently short petioles. So far known only from the middle part of the Cape York Peninsula.

Croton byrnesii *Airy Shaw* in Muelleria 4: 225 (1980). Type: N.T., Cannon Hill, 1972, *Byrnes* 2833 (DNA).

NT (DG)—Endemic.

Shrub or small slender tree of 2–4 m, in monsoon rain-forest, or in broad-leaf scrub in sandy soil at base of sandstone outcrop, or at edge of creek with *Tristania*, at low altitudes.

Distinguished from the closely related *C. armstrongii* but its almost total glabrescence, in its often long-stipitate petiolar glands, and in the greater distance separating these from the base of the lamina.

A critical micro-species from the Cannon Hill area.

Croton capitis-york *Airy Shaw* in Muelleria 4: 226 (1980). Type: Q., Temple Bay Yards (Cape York Penins.), 1976, *Hyland* 8971 (QRS).

Q (CK)—Endemic.

Shrub to 4 m tall, in dry monsoon forest on stony red soil, or in evergreen vine thicket on fine white sand with dark stained topsoil, at 70–150 m. Fig. 2A.

Differs from *C. argyratus* and *C. insularis* in the conspicuous stipitate glands at the apex of the petiole, and in the greenish rather than pinkish colour of the lepidote undersurface on drying. The leaves are considerably larger and thinner than those of *C. insularis*, whilst the female sepals are very much smaller than those of *C. argyratus*. The flowers of *C. capitis-york* are so far known only in the bud stage, but it seems probable that the style-branches will be found to be twice bifid, rather than simple.

The species is a distinct one, only known from the northern half of the Cape York Peninsula.

Croton densivestitus *White & Francis* in Proc. Roy. Soc. Queensl. 35: 80–83, fig. 9 (1923); Airy Shaw in K.B. 31: 385 (1976). Type: Q., Harvey's Creek, 1889, *F. M. Bailey* s.n. (BRI).

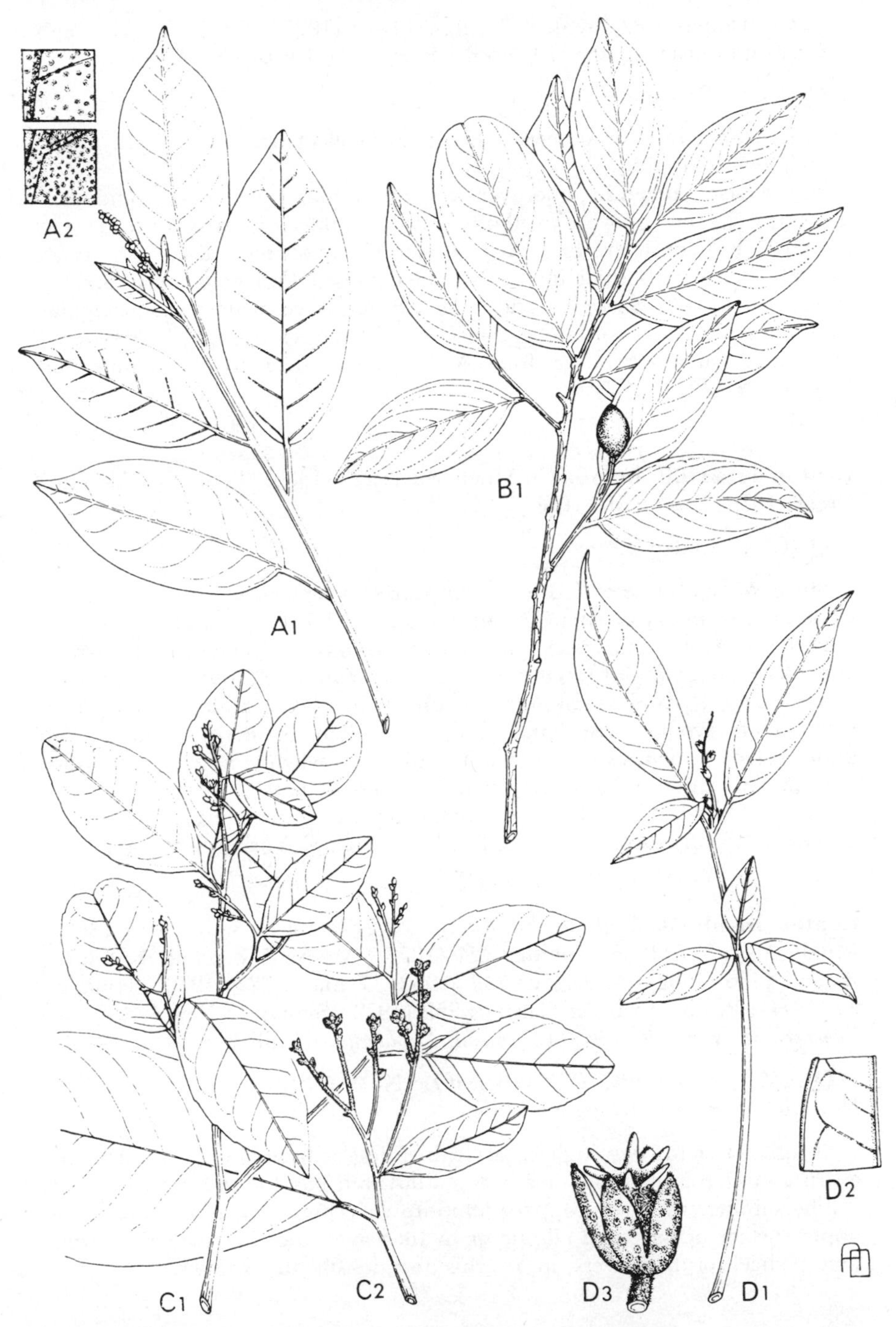

FIG. 2. *Croton capitis-york:* **A1** habit × $\frac{1}{3}$, from *Hyland* 9021; **A2** leaf surfaces × 2, from *Hyland* 9021. *Drypetes subcubica:* **B1** habit × $\frac{1}{3}$, from *Irvine* 1454. *Dimorphocalyx australiensis:* **C1** habit × $\frac{1}{3}$, from *Smith* 11841; **C2** inflorescence × $\frac{8}{9}$ from *Smith* 14357. *Croton dockrillii:* **D1** habit × $\frac{1}{3}$, from *Webb & Tracey* 6083; **D2** lower leaf surface × $\frac{2}{3}$, from *Webb & Tracey* 6083 **D3** flower × 6, from *Webb & Tracey* 6083.

C. pubens Domin: 882 [328], t. 31, figs. 11–19 (1927). Type: Q., Harvey's Creek and near estuary of Russell River, 1909–10, *Domin s.n.* (PR).

Q (CK)—Endemic.

Tall shrub in rain-forest below 100 m; once collected on Mt Bellenden-Ker at unknown altitude.

The coarse, spreading, more or less ochraceous pubescence, bifid (not 4–6-fid) styles, and obovate to elliptic leaves, rounded or narrowly cordate at the base, distinguish this local species from *C. stigmatosus* and *C. phebalioides*, the only species with which it is likely to be confused. Other stellate-pubescent species, such as *C. arnhemicus* and *C. tomentellus*, have ovate to suborbicular leaves, often broadly cordate at the base. The species seems to be restricted to a small area extending from Mt Bellenden-Ker to the Palmerston Highway.

Croton dockrillii *Airy Shaw* in Muelleria 4: 227 (1980). Type: Q., Alligator Creek, 1972, *Dockrill* 589 (QRS).

Q (CK)—Endemic.

Shrub or small tree of 2–3 m, in rain-forest near beach or in littoral vine woodland, or in dry or riparian rain-forest, at 20–75 m. Fig. 2D.

Perhaps nearest to *C. armstrongii* and *C. byrnesii*, but the leaves are narrower and more elongate and often conspicuously acuminate, and the margin is more distantly and shallowly repand-denticulate, and conspicuously reflexed or revolute. The basal glands are often distinctly stipitate. The undersurface, which is sparsely stellate-pubescent when young, is seen under high magnification to be densely minutely puncticulate.

The revolute leaf-margins give this species a very distinctive appearance, at least in the dried state. It is confined to the Cape York Peninsula.

Croton insularis *Baill.* in Adansonia 2: 217 (1862) & 6: 301 (1866); Muell. Arg.: 527 (1866); Benth.: 124 (1873); Moore & Betche: 76 (1893); Bailey: 1436 (1902); Anderson: 218 (1968); Francis: 227 (1970); Hyland: 30 (1971); Beadle, Evans & Carolin: 252 (1972). Syntypes: New Caledonia, *Pancher* 360, *Vieillard* 1136, 1137, 1138, *Deplanche* 483 (all P).

Q (CK, NK, LT, PC, BT, WB, MO); **NSW** (NC, CC, NT, CT, ST)—New Hebrides, New Caledonia.

Shrub or tree to 10 m high, in dry rain-forest or brigalow-softwood scrub, on rocky outcrops or brown loam or shallow red basaltic soil, up to 750 m.

The shining silvery or coppery lepidote covering of all parts except the upper surface of the small elliptic or ovate leaves, together with the rather long pedicels of the flowers, makes this an unmistakable species.

Croton magneticus *Airy Shaw* in Muelleria 4: 227 (1980). Type: Q., Magnetic Island, 1938, *Goy* 329 (BRI).

Q (NK)—Endemic.

Straggling shrub along rocky seashore; only known from Magnetic Island.

Closely related to *C. pilophorus* Airy Shaw, of eastern New Guinea, but differing in its smaller leaves, dense-flowered inflorescences and linear style-branches. The two species are distinguished from all others in the region by their scaberulous indumentum, by their shallowly indentate-crenate leaf margin, and by the dull muddy-brown colour assumed on drying.

CROTON OPPONENS *F. Muell. ex Benth.*: 125 (1873); Bailey: 1436 (1902).

Q. 'The collector's name and precise station not given. The species requires further elucidation from more perfect specimens' (Benth., *l.c.*). I have seen the type specimen in MEL. It is not referable to the genus *Croton*, and probably not even to the family *Euphorbiaceae*. I am not able to suggest an identification, but the plant should be recognizable by someone familiar with the Queensland flora.

Croton phebalioides *F. Muell. ex Muell. Arg.* in Flora 47: 485 (Oct. 1864) & in DC.: 581 (1866); Baill. in Adansonia 6: 301 (1866); Benth.: 125 (1873), *p.p.*; Bailey: 1436 (1902): Moore & Betche: 76 (1893); Baker, Hardw. Austr.: 358 (1919); Anderson: 218, 377 (1968); Francis: 227 (1970); Airy Shaw in K.B. 31: 386 (1976). Type: Q., Burdekin River, *F. Mueller* in Herb. Hook. (K).

C. maidenii R. T. Baker in Journ. & Proc. Roy. Soc. N.S.W. 48: 444, t. 12 (1915). Type: N.S.W., Guthrie's Mountain (Read's Mine), *Paddison s.n.* (NSW).

C. phebalioides var. *acuminatus* Domin: 880 [326], fig. 150, *sinistr.* (1927). Syntypes: Q., sine loc., *Dietrich* 2326 (PR); Edgecombe Bay, *Dallachy* (K).

Q (NK, SK, PC, LT, MI, WG, MN, MO); **NSW** (NC, NWS)—Endemic.

Shrub or small tree to 4 m high, in shallow sandy soil in mixed woodland or on sandy or rocky slopes or softwood scrub or on rocky hilltops or in dry rain-forest, up to 390 m.

Croton phebalioides is the dry-country counterpart of the moist rain-forest *C. stigmatosus*. The leaves are usually much smaller and narrower, and entire or almost so.

Croton cf. **prunifolius** *Airy Shaw* in K.B. 33: 56 (1978). Type: W New Guinea, *Kostermans & Soegeng* 275 (holotype K).

WA (K)—Lesser Sunda Is., New Guinea.

No ecological information for Australia. [In Papua a shrub or tree to 12 m tall, at edge of mangrove swamp, or in dry rain-forest/savanna edge, occasionally in forest, at 0–660 m.]

Sparsely stellate-pubescent or glabrescent, with rather small, elliptic, very *Prunus*-like leaves, 5–10 × 2–3·5 cm, inconspicuously serrulate, chartaceous; petiole very slender, up to 3 cm long; inflorescences slender, to 10 cm long, ♂ and ♀ flowers intermixed; capsules closely stellate-lepidote.

A single sterile specimen at BRI closely resembles the New Guinea material, but further and fertile gatherings will be necessary for certainty.

Croton stigmatosus *F. Muell.*, Fragm. 4: 140 (Nov. 1864); Muell. Arg. in Linnaea 34: 107 (1865) & in DC.: 580 (1866); Baill. in Adansonia 6: 301 (1866). Syntypes: N.S.W., Richmond River, *Beckler*; Q., Moreton Bay, *Leichhardt*, *F. Mueller*; Broad Sound, *Bowman*; Fitzroy River & Mt Mueller, *Dallachy* (all MEL).

C. stigmatosus var. *eurybioides* Baill., *l.c.* (1866), *in obs.* Material not specified.
C. phebalioides var. *hirsuta* Bailey: 1436 (1902). Type: Q., Taylor's Range, near Brisbane, *Bailey* (BRI).
C. phebalioides var. *stigmatosus* (F. Muell.) Domin: 880 [326], fig. 150, *dextr.* (1927).

Q (PC, MO); **NSW** (NC)—Endemic.

Shrub or tree to 10 m high, in rain-forest on basalt up to 450 m.

Croton stigmatosus approaches very closely to *C. phebalioides*, but probably deserves specific recognition. It appears to represent the broad-leaved or large-leaved rain-forest counterpart of that species. The leaf-margin is usually clearly (or sometimes minutely) repand-denticulate.

Croton stockeri (*Airy Shaw*) *Airy Shaw* comb. & stat. nov., *cum descr. amplif.*

Croton wassi-kussae Croiz. var. *stockeri* Airy Shaw in Muelleria 4: 229 (1980). Type: Q., between Rocky River and Massey Creek, 1973, *Stocker* 1076 (QRS).

Q (CK)—Endemic.

Shrub of 2·5 m in dry rain-forest on old sand-dune at 80 m.

Branchlets terete, closely minutely stellate-pubescent, but the stellate hairs masked by the erect central rays which form a dense brownish-ochraceous tomentum. Leaves ovate, up to 7·5 × 4·5 cm, shortly cordate at the base, narrowed to an acute apex, margin entire or very obscurely denticulate, firmly chartaceous, dark brown when dry, dull on both surfaces, the upper surface rather closely minutely stellate with occasional central rays, the lower surface densely brown-stellate with the erect central rays forming a dense tomentum; midrib and 6–7 pairs of primary nerves distinctly incised above; basal glands discoid, sessile; petiole 0·5–2 cm long, slender, tomentose; stipules obsolete. Inflorescence to 8 cm long; male and female flowers mixed, the males greatly preponderating but still in very young bud, the few females very proterogynous. Calyx of female flower 6–7 mm diam.; segments triangular, acute, brown-stellate and pallid-pilose; pedicel 3–4 mm long. Petals 0, or rarely represented by a short white thread. Disk low, ± annular. Ovary subglobose, 4 mm diam., muricate, each tubercle crowned by a many-rayed stellate hair, and densely pilose in addition. Styles 3, spreading, adpressed to the ovary, flat and ribbon-like, glabrous, 1·5 mm long, either simple, truncate and very shortly bifid at the apex, or bearing on each side just before the apex a pair of lateral branches 1·5 mm long, the proximal one of each pair directed backwards (retrorse) and the distal ± forwards; each branch is thickened or bifid at the tip. Capsule unknown.

I have come to the conclusion that there are too many divergent characters to warrant maintaining this distinct plant as a mere variety of the Papuan *C. wassi-kussae*. Particularly the strongly developed tomentose indumentum, together with the muricate ovary and either simple or curiously branched style-arms, are strikingly different from the characters of the Papuan plant. There is perhaps also a more distant relationship with *C. magneticus* and *C. pilophorus* (*vide* p. 620, *supra*). I am therefore raising var. *stockeri* to specific rank and providing a full description.

The type of branching of the style-arms is remarkable. I have seen nothing comparable in any other Asiatic or Australian species, nor has my colleague Mr A. Radcliffe-Smith seen it in any African species. Further material of this plant would be welcome.

Croton cf. **storckii** (*Muell. Arg.*) *A. C. Sm.* in Bull. B.P. Bishop Mus. 141: 83 (1936). Type: Fiji, *Storck* (G–DC).

C. storckii Seem. in Bonplandia 10: 297 (1862), *nomen.*
C. verreauxii var. *storckii* Muell. Arg. in Linnaea 34: 117 (1865) & in DC.: 621 (1866); Seem., Fl. Vit.: 222, t. 57 (1867).

Q (CK)—Fiji.

Shrub or small tree in open forest at 75 m.

A sterile branchlet from a bush growing by the Claudie River in the Cape York Peninsula, 75 m, 17 Oct. 1974, *Hyland* 7817, is remarkably like the Fijian *C. storckii*, and could well prove to be that species, but fertile material is needed. The plant is almost completely glabrous; the leaves elliptic or obovate, up to 11 × 5 cm, very smooth and green when dry; petiole very slender, up to 3·5 cm.

Croton tomentellus *F. Muell.*, Fragm. 4: 141 (1864); Muell. Arg. in Linnaea 34: 108 (1865) & in DC.: 591 (1866); Baill. in Adansonia 6: 300 (1866); Benth.: 126 (1873); Ewart & Davies: 167 (1917); Gardner: 72 (1931); Airy Shaw in Muelleria 4: 228, 229 (1980). Type: N.T., Victoria River, *F. Mueller* (MEL).

WA (K); **NT** (DG, VR)—Java.

Shrub to 4·5 m tall, in tall open forest at edge of mangrove in red lateritic soil, or among granite boulders, at low altitude.

Near *C. arnhemicus*, but differing in its subentire or finely crenulate leaves, more weakly 3–5-nerved at the base, only ± 10 stamens, and thinly stellate-pilose capsule.

Sterile indeterminable gatherings of plants apparently related to *C. tomentellus* have been seen from the Kimberleys, Western Australia (open woodland on basaltic loam on slope, Prince Regent River Reserve), and from the Burke District of Queensland (fragmented deciduous vine thicket on limestone outcrop, near Lawn Hill Homestead).

Croton triacros *F. Muell.*, Fragm. 6: 185 (1868); Benth.: 127 (1873); Bailey: 1437 (1902); Hyland: 67 (1971). Type: Q., Rockingham's Bay, *Dallachy* (MEL).

Q (CK, NK)—Endemic.

Shrub or small tree to 3 m tall, in open *Eucalyptus* forest at unknown altitude.

Differs from the closely related *C. brachypus*, of the Cape York Peninsula, in the much longer petiole and less cuneate lamina of the leaves. In the dried state the leaves have the same smooth oily green appearance as those of *C. brachypus*. (The leaves of *Glochidion sessiliflorum* have a similar appearance.) In both *C. triacros* and *C. brachypus* the leaves are practically glabrous, and entire or obscurely crenate or serrate, and the venation is slender, ± immersed and inconspicuous.

The species is known from an area extending from the Daintree River south to Rockingham Bay.

Croton verreauxii *Baill.*, Ét. Gén. Euphorb.: 357, Atlas: 32, t. 17, fig. 10 (1858); F. Muell., Fragm. 4: 141 (1864); Muell. Arg. in Linnaea 34: 117 (1865) & in DC.: 620 (1866); Baill. in Adansonia 6: 302 (1866); Benth.: 126 (1873); Moore & Betche: 76 (1893); Bailey: 1437 (1902); Baker, Hardw. Austr.: 358 (1919); Anderson: 218, 377 (1968); Francis: 227 (1970); Beadle, Evans & Carolin: 252 (1972); Airy Shaw in K.B. 31: 387 (1976) (*excl. synon.*). Type: N.S.W., 'Camp in Heaven', *Verreaux* 59 or 159 (P).

Q (MO); **NSW** (NC, CC, SC)—Endemic.

Shrub or small tree to 8 m tall, fairly frequent on steep hills with eucalypts on the slopes and broad-leafed forest in the valleys, or on rain-forest margins or in fringing forest on basalt, up to 540 m.

Croton verreauxii, like *Glochidion ferdinandii*, is a species that has been much misunderstood and too widely interpreted in the past. One of its characteristic features is the dark purplish pigmentation of its young stems and petioles; a purplish tinge is often clearly seen also on the undersurface of the leaves. Whilst this coloration is not invariably present (it is evident in at least 75 per cent of the herbarium specimens that I have examined), I have seen it in no other Australian *Croton*, and it is therefore a useful indication of identity.

Except for the young growth and inflorescences the plant is almost glabrous. The leaf-margins are usually distinctly crenate-serrulate.

The distribution of *C. verreauxii* is restricted to a strip of country running from southeast Queensland to southeast New South Wales.

Dimorphocalyx *Thw.*

Distinguished from the related *Trigonostemon* by the leaves being not or obscurely triplinerved at the base, glandular-denticulate to entire or occasionally remotely crenate; the inflorescences always axillary; the flowers usually dioecious: the petals usually white; the stamens 7–15, the outer ± free, the inner united into a column; and the calyx frequently accrescent in fruit.

Dimorphocalyx australiensis *C. T. White* in Proc. Roy. Soc. Queensl. 47: 80 (1936); Airy Shaw in K.B. 23: 125 (1969) & 29: 328 (1974). Type: Q., Mowbray River, *Brass* 2019 (BRI).

Tritaxis australiensis S. Moore in Journ. Linn. Soc., Bot., 45: 218 (1920). Type: Q., Cape York, *Daemel s.n.* (BM).

Q (CK)—Papua, Lesser Sunda Is.

Shrub or small tree to 10 m high, in rain-forest or gallery forest or monsoon forest, or in scrub, at 60–500 m. Fig. 2C.

This is the only eastern representative of an otherwise entirely Indian and West Malesian genus. *Dimorphocalyx* differs from the related *Trigonostemon* in that the leaves are rarely triplinerved at the base, the inflorescences always axillary, the flowers usually dioecious, with white petals, and the calyx frequently accrescent in fruit. It is also related to *Hylandia*, but differs in its much smaller stature, smaller, thin, usually crenate or dentate leaves, axillary inflorescences, few stamens, pilose filaments and trilocular ovary and capsule.

Dimorphocalyx australiensis occurs in scattered localities from the Mowbray River (type locality, ± 15 km S of Mossman) to Cape York, and again in southern Papua and in some of the Lesser Sunda Islands.

Dissiliaria *F. Muell. ex Benth.*

Trees or shrubs, with opposite, entire or crenulate, coriaceous or chartaceous leaves; stipules deltoid or shortly oblong, interpetiolar, very quickly caducous. Flowers in small axillary cymes. Male flower: sepals 4–6, imbricate; stamens 15–20, very short, from a pubescent receptacle, sometimes interspersed with small disk-glands; pistillode 0. Female flower: sepals 3 + 3; disk annular or shortly cupular; ovary 3(–4)-locular, loculi 2-ovulate; styles 3, simple, shortly connate, recurved. Capsule 3(–4)-locular.

Good flowering material (both ♂ and ♀) of all three species is much needed.

1. Leaves small, up to 5 × 3·7 cm, chartaceous, broad-ovate or rhombic-ovate, closely crenulate to subentire . . . . . . **D. muelleri**
1. Leaves larger, up to 18 × 7 cm, coriaceous, ovate, obovate or elliptic, entire:
  2. Nerves close, only slightly elevated beneath . . . **D. baloghioides**
  2. Nerves lax, very distinctly elevated beneath . . . . **D. laxinervis**

Dissiliaria baloghioides *F. Muell.* [ex Baill. in Adansonia 7: 359 (1867), *gen. nondum definito nec rite descripto*] *ex Benth.*: 90 (1873); Bailey: 1430 (1902); Baker, Hardw. Austr.: 359 (1919); Pax & Hoffm. xv: 292 (1922); Francis: 219, figs. 131, 132 (1970), *pro parte*. Type: Q., Moreton Bay, *Cunningham* (K).

Q (PC, MO)—Endemic.

Tree to 36 m high, in rain-forest up to 500 m altitude.

Glabrous except for young branchlets and inflorescences. Leaves opposite, ovate, obovate or elliptic, up to 12 × 5 cm, broadly narrowed into the petiole, entire, obtuse, very smooth and shining above, finely and 'smoothly' reticulate below (as in *Drypetes* spp.), brown when dry, thinly coriaceous; petiole slender, 5–8 mm long. Male flowers in small crowded cymes in distal or subdistal leaf-axils. Sepals 3 + 3, elliptic, 2·5 mm long; stamens 15–20, about 2·5 mm long, arising from a pubescent receptacle. Female flowers solitary in distal leaf-axils; pedicel 4–5 mm long; ovary 2·5 mm diam., pubescent; styles 3, shortly connate, strongly recurved. Capsule globose, about 1·8 mm diam., woody, minutely ochraceous-tomentellous; seeds not seen.

Good flowering material of this species seems to be very rarely collected. The tree is confined to the SE of Queensland, from Sarina to the Brisbane region.

Dissiliaria laxinervis *Airy Shaw* in Muelleria 4: 220 (1980). Type: Q., Claudie River, *Hyland* 2578 RFK (holotype K).

Dissiliaria sp. aff. *D. baloghioides*, Hyland: 85 (1971).

Q (CK)—Endemic.

Tree to 25 m tall in gallery rain-forest or on granite wash at 30–80 m.

Closely related to *D. baloghioides*, but differing in the much laxer nervation, which is clearly elevated on the leaf-undersurface. The fruiting pedicels are sometimes 4·5 cm long. In the type form, from the Claudie River (12° 45′ S), the leaves are cuneately narrowed to the base, and have remained green or greenish on drying, but in a specimen from Gap Creek (15° 45′ S) the leaves are rounded at the base and are reddish brown when dry. Flowers of this species have not yet been collected.

Dissiliaria muelleri *Baill.* [in Adansonia 7: 359, t. 1 (1867), *gen. nondum definito nec rite descr.*] *ex Benth.*: 91 (1873); Bailey: 1430 (1902); Pax & Hoffm. xv: 292 (1922); Francis: 226 (1970). Syntypes: Q., Rockhampton, *Thozet*, *Dallachy* (P, MEL).

Q (PC, WB)—Endemic.

Small tree in dry rain-forest at 100 m.

Glabrous except for very young growth. Branchlets slender, 'twiggy', often rough with small lenticels. Leaves opposite, small, broad-ovate or rhombic-ovate, up to 5 × 3·7 cm, cuneate at the base, obtuse or rounded or occasionally retuse at the apex, margin very shallowly but closely crenulate to subentire, firmly chartaceous, finely reticulate-venulose, especially below, very shortly and slenderly petiolate. Male inflorescences very short and few-flowered, sometimes reduced to a single flower. Sepals 4–5, broad-ovate or suborbicular, membranaceous. Stamens ± 20, very short, from a pilose receptacle, with numerous small disk-glands interspersed. Female inflorescences 1–3-flowered, up to 3 cm long, pedicels about 1 cm long. Sepals 3 + 3, broadly elliptic, 2–3 mm long. Disk annular or shortly cupular, crenulate, surrounding the base of the ovary. Ovary very shortly whitish-puberulous, 2 mm diam.; styles shortly connate, subulate, recurved. Capsule 3(–4)-locular, 10–15 mm diam.

A scarce species, only known from near Rockhampton and the region of Mt Bauple, north of Gympie. Flowering and fruiting seems to be rarely observed at the present day.

A variety or related species, with narrowly elliptic leaves up to 4·5 × 1·8 cm, has been collected at Rocky River in the Cape York Peninsula (*Hyland* 2554 RFK).

Drypetes *Vahl*

Trees or shrubs, occasionally rheophytic. Leaves alternate, shortly petioled, penninerved, entire or sometimes dentate, often unequal-sided at the base; stipules various, small or large, occasionally pectinate, mostly caducous. Flowers dioecious, axillary or cauliflorous, fascicled. Male flower: sepals 4–5, free, suborbicular, much imbricate; stamens 3–50, free, anthers mostly relatively large; disk central, disciform, flat or with variously raised or sinuate or lobulate margin, the lobes sometimes produced outward between the filaments or embracing their base. Female flower: sepals much as in male; disk annular; ovary 1–3-locular, loculi biovulate; styles very short, or occasionally more elongate; stigmas flabellate, rarely bifid with linear lobes. Drupe globose or flattened-ellipsoid or transversely sub-bilobed, ± coriaceous or crustaceous; endocarp occasionally sculptured; seed(s) ± conformable to cavity of endocarp.

1. Leaves up to 20 cm long, distinctly acuminate; venation coarse and lax, prominently elevate; ovary 1-locular, pubescent; fruit rounded-subcubic **D. subcubica**
1. Leaves up to 10 cm long, not or scarcely acuminate; venation fine and close, scarcely elevate; ovary 2-locular; fruit ovoid . . . **D. lasiogyna**
   2. Ovary and fruit pubescent . . . . var. **lasiogyna**
   2. Ovary and fruit glabrous . . . . var. **australasica**

Drypetes lasiogyna (*F. Muell.*) *Pax & Hoffm.* xv: 272 (1922); Airy Shaw in K.B. 31: 364 (1976). Type: N.T., Port Essington, *Leichhardt* (MEL).

var. **lasiogyna:** ovary and fruit pubescent.

Hemicyclia lasiogyna F. Muell., Fragm. 4: 119 (1864); Baill. in Adansonia 6: 330 (1866); Benth.: 118 (1873); Ewart & Davies: 165 (1917); Chippendale in Proc. Linn. Soc. NSW 96: 244 (1972).
H. sepiaria var.? *oblongifolia* Benth.: 117 (1873). Type: N.T., Port Darwin, on the beach, *Schultz* 746 (K).
[*Drypetes australasica* sec. Pax & Hoffm. xv: 271 (1922), quoad synon. '*Hemicycliam sepiariam* Benth.' et specim. 'Port Darwin (Schultz n. 746!)' tantum; Chippendale, *l.c.* (1972); *non* (Muell. Arg.) Pax & Hoffm.]
[*Hemicyclia australasica* sec. Maiden, For. Fl. NSW 8 (74): 57, tab. phot. (1923), *vix* Muell. Arg.]
[*Hemicyclia sepiaria* sec. Chippendale, *l.c.* (1972), *non* Wight & Arn.]

NT (DG); **Q** (CK)—New Guinea.

Shrub or tree to 8 m high, in forest above beach, or on creek bank in sandstone gorge, or in monsoonal rain-forest on gravelly lateritic soil, or in rain-forest community within woodland, at low altitudes.

The type form of the species, with pubescent ovary, is almost confined to the Northern Territory, but in recent years has been unexpectedly found on the Atherton Tableland (R. 194 Western, 17° 20′ S, 145° 25′ E, SW of Atherton, n.d., *Volck* 4386). There is great variation in the size and shape of the leaves and of the fruit.

var. **australasica** (*Muell. Arg.*) *Airy Shaw* comb. nov.

[*Hemicyclia sepiaria sec.* F. Muell., Fragm. 4: 119 (1864) & 6: 182 (1868); Baill. in Adansonia 6: 330 (1866); *non* Wight & Arn.]
H. australasica Muell. Arg.: 487 (1866); Benth.: 118 (1873); Moore & Betche: 75 (1893); Baker, Hardw. Austr.: 355 (1919); Maiden, For. Fl. NSW 8(74): 55, t. 282 (1923) [excl. tab. phot. p. 57]; Francis: 219, figs. 129, 130 (1951). Type: Q., Burdekin River, granite rocks, *F. Mueller* (G–DC, K).
H. sepiaria var. *australasica* (Muell. Arg.) Baill. l.c., *in obs.* (1866).
Drypetes australasica (Muell. Arg.) Pax & Hoffm. xv: 270 (1922); Anderson: 216 (1968); Francis: 219, figs. 129, 130 (1970); Hyland: 48 (1971); Airy Shaw in K.B. 31: 364 (1976).

Ovary and fruit glabrous.

Q (CK, NK, PC, WB, MO); **NSW** (NC)—New Guinea.

Shrub or 'fair-sized tree', in rain-forest on granite or granodiorite, or in Araucarian heath, or in littoral forest bordering mangrove swamps, or on sand-shingle islands, up to 1200 m.

With the breakdown noted above in what was formerly believed to be the strictly vicarious occurrence of *D. lasiogyna* in the Northern Territory and *D. australasica* in Queensland, it no longer seems justifiable to treat these two closely related taxa as species. I therefore reduce *australasica*, with a glabrous ovary, to a variety of the earlier described *D. lasiogyna*. Var. *australasica* has not yet been found in the Northern Territory.

Drypetes subcubica (*J. J. Sm.*) *Pax & Hoffm.* xv: 250 (1922); Backer & Bakh. f.: 472–3 (1963); Airy Shaw in K.B. 29: 300 (1974) & 31: 365 (1976). Syntypes: Java, *Koorders* 2140, 2145, 2148, 26179 (BO); *Koorders* 25289, 37987 (BO, K).

Cyclostemon subcubicus J. J. Sm. in Koord. & Valet., Bijdr. No. 12 Booms. Java, in Meded. Dep. Landb. 10: 216 (1910).
Drypetes sp. (= AFO/780), Hyland: 64 (1971).

Q (CK)—Java, Lesser Sunda Is.

Tree to 12 m high, in rain-forest at 760 m. Fig. 2B.

The far coarser and laxer, prominently raised venation of the mostly larger (to 20 cm long) and more distinctly acuminate leaves of *D. subcubica* distinguish it at a first glance from *D. lasiogyna*. The stamens of the short-pedicelled ♂ flower are about 6 in number. The ovary is bilocular (§ *Sphragidia*) and pubescent, and the fruit (not yet collected in Australia) is rounded-subcubic, with 4 shallow longitudinal grooves.

The species is so far known in Australia only from a limited area north of Cairns.

Endospermum *Benth.*

Trees, sometimes of considerable height, with a thin stellate indumentum or glabrescent, twigs conspicuously marked with large round scars of fallen leaves, sometimes hollow and myrmecophilous. Leaves crowded towards tips of branches, ovate or orbicular, often peltate or cordate, entire, coriaceous, penninerved, but palmately 3–8-nerved at the base, often with a conspicuous pale globose gland at each main nerve-junction; petiole longish, biglandular at the apex; stipules small, ovate, very early caducous. Flowers dioecious. Male inflorescences paniculate, many-flowered, sometimes long-peduncled; bracts acute. Calyx shortly cupular, very shortly 3–4-dentate; extra-staminal disk 4–5-angled, lobulate; stamens 5–10, ± connate, anthers 4-locellate; pistillode 0. Female inflorescence simple or sometimes weakly paniculate, mostly distinctly peduncled. Calyx 5-toothed; disk annular; ovary globose, 2- or 4–6-locular, styles connate into a sessile disc. Fruit a ± fleshy indehiscent capsule or drupe; seed subglobose or flattened, smooth, ochraceous.

Endospermum myrmecophilum *L. S. Smith* in Proc. Roy. Soc. Queensl. 58: 56, t. II (1947); Schaeffer in Blumea 19: 181, 187, map 3 (1971); Hyland: 66 (1971); Airy Shaw in Muelleria 4: 228 (1980). Type: NE New Guinea, Yalu, 1944, *White, Dadswell & Smith* NGF 1640 (holotype BRI).

Q (CK, NK)—New Guinea.

Tree to 36 m high (in New Guinea), in rain-forest at low altitude.

Closely related to *E. medullosum* L. S. Smith, of New Guinea, and apparently only distinguishable from it with reasonable certainty by the usually much weaker indumentum and laxer and less prominent venation of the leaf-undersurface (cf. L. S. Smith, *l.c.*: 53, sub *E. medullosum*). There is also a tendency for *E. myrmecophilum* to have hollow branchlets, peltate leaves and less prominent petiolar glands, but these features seem inconstant and unreliable.

Only two Australian collections have been seen, from widely separated localities (Claudie River, *Hyland* 3101 RFK, with solid stem; Mission Beach, *Hyland* 02050, with hollow stem), with identical nervation and indumentum, on the strength of which I refer both to *E. myrmecophilum*.

Euphorbia *L.*

(incl. *Chamaesyce* (Rafin.) S. F. Gray)

Annual, biennial or perennial herbs, suffrutices, shrubs or trees, sometimes of cactoid habit, unarmed or aculeate, with a milky latex. Leaves alternate, opposite or verticillate, simple, entire, toothed or lobed, uniform or heterophyllous, membranaceous, chartaceous or coriaceous, penninerved or palminerved, petiolate or sessile, stipulate or not. Stipules, when present, filiform or triangular and entire, toothed or laciniate, commonly interpetiolarly connate. Flowers commonly monoecious, with one female flower and several male flowers enclosed together within a small cupuliform glanduliferous involucre termed a 'cyathium', the whole structure functioning as a floral unit—i.e. pseudanthially. Cyathia axillary and solitary or else disposed in lax or dense pseudodichasia or pseudopleiochasia. Cyathial involucre usually 5-lobed, with (1–)4–5 variously-shaped appendiculate or exappendiculate glands alternating with the lobes; one interlobular sinus larger than the rest except when 5 glands are present. Male flowers bracteate or ebracteate, achlamydeous, each consisting of a single pedicellate stamen, with a well-marked articulation between pedicel and filament; anthers biglobose, vertically, obliquely or subhorizontally dehiscent. Female flower consisting of a pedicellate subachlamydeous trilocular ovary; pedicel commonly elongating and becoming recurved through the largest interlobular sinus of the cyathial involucre and then later straightening, in fruit. Ovules 1 per loculus. Styles 3, united at the base or for up to $\frac{2}{3}$ their length, usually bifid, rarely undivided. Fruit a smooth, ridged or tuberculate regma, dehiscing at first septicidally and septifragally, leaving a persistent columella, and then the valves almost immediately partially dehiscing loculicidally; exocarp chartaceous, mesocarp usually thin, endocarp ligneous. Seeds varying greatly in shape, size and colour, smooth, sculptured or variously ornamented, with a ventral raphe and commonly with a chalazal truncation or depression and a paramicropylar caruncula. (A. Radcliffe-Smith.)

In Australia the genus *Euphorbia* in the wide sense comprises some 45 species, made up of *Euphorbia s. str.* 7 spp. and *Chamaesyce* 38 spp. (see Hassall in *Austr. J. Bot.* 24: 633–640 (1976) & 25: 429–453 (1977)). A revision of the Australian species of *Euphorbia s. str.* is presented by Hassall (1977, *l.c.*). A revision of the members of the *Chamaesyce* group has yet to be undertaken.

Excoecaria *L.*

Trees or shrubs, glabrous, mostly dioecious. Leaves alternate or opposite, penninerved, entire or crenulate, shortly petioled; stipules minute. Inflorescences axillary or terminal, the males or male portions densely spiciform, the females laxer and fewer-flowered, or in androgynous inflorescences represented by 1–few female flowers at the base of the males; bracts small, of various shape, usually accompanied by 2 lateral glands at the base. Male flower: sepals 2–3, free or shortly connate, imbricate, entire or toothed; disk 0; stamens 2–3, free; pistillode 0. Female flower: sepals 3, free or shortly connate; ovary 3-locular, loculi 1-ovulate, styles simple, shortly connate, recurved. Capsule 3-locular, dehiscent, sometimes with a persistent columella, sometimes with a ± fleshy outer layer; seeds globose or oblong, mostly ecarunculate.

1. Leaves narrow-oblong, spathulate, or cuneate-obovate, entire, borne in sessile tufts or on short brachyblasts; stamens 2 . . . . **E. parvifolia**
1. Leaves elliptic or ovate, shallowly crenate-serrulate or subentire, borne normally on the main branches; stamens 3:
  2. Mangrove shrub or tree; leaves often subentire, apex rounded or obtuse or occasionally very shortly subacuminate . . . . **E. agallocha**
  2. Tree of coastal dry rain-forest; leaves rarely subentire, apex distinctly subacuminate-attenuate . . . . **E. dallachyana**

Excoecaria agallocha *L.*, Syst., ed. 10, 2: 1288 (1759) & Sp. Pl., ed. 2: 1451 (1763); Muell. Arg.: 1220 (1866); Baill., Ét. Gén. Euphorb., Atlas: t. 7, figs. 31–34 (1858) & in Adansonia 6: 324 (1866); Benth.: 152 (1873); Moore & Betche: 79 (1893); Bailey: 1456 (1902); J. J. Sm.: 616 (1910); Pax & Hoffm. v: 165 (1912); Gardner: 72 (1931): Backer & Bakh. f.: 499 (1963); Anderson: 220 (1968); Airy Shaw in K.B. 26: 268 (1972) & K.B. Add. Ser. IV: 112 (1975). Type: Amboina, *Rumphius*, Herb. Amboin. 2: 237, tt. 79, 80 (1741) (*Arbor excoecans* Rumph.).

E. affinis Endl., Prodr. Fl. Norfolk.: 83 (1833). Type: Q. [PC], 'in Novae Hollandiae orientalis sinu Broad-Sound dicto', *Ferd. Bauer* (W?).
E. ovalis Endl., Prodr. Fl. Norfolk.: 83 (1833). Type: Q.[BK], 'in Novae Hollandiae septemtrionalis sinu Carpentaria', *Ferd. Bauer* (W?).
E. agallocha var. *ovalis* (Endl.) Muell. Arg.: 1221 (1866); Pax & Hoffm. v: 167 (1912).

WA (K); **NT** (DG, VR); **Q** (BK, CK, NK, PC, WB, MO); **NSW** (Far NC)—S India and Ceylon to Formosa and Ryu-Kyu Is., and throughout Malesia to Pacific.

Shrub or tree to 8 m tall, in mangrove formations, at contact of mangrove and rock platform, or at junction of shingle ridge and swamp, or along margins of shingle ridges and mangrove woodland, or on clayey mud on tidal flats, or occasionally in littoral forest on sandy soil, at or just above sea-level.

Excoecaria dallachyana (*Baill.*) *Benth.*: 153 (1873); Bailey in Queensl. Agric. Journ. 3: 284, t. 54 (1898) & Queensl. Fl. 5: 1457 (1902); Pax & Hoffm. v: 168 (1912); Baker, Hardw. Austr.: 363 (1919); Anderson: 220 (1968); Francis: 234 (1970). Syntypes: Q., Rockingham Bay?, 1863, *Dallachy* 248 (MEL); sine loc., *Bowman* 162 (MEL).

E. agallocha var. *dallachyana* Baill. in Adansonia 6: 324 (1866).

Q (PC, LT, MO); **NSW** (Far NC)—Endemic.

Small tree to 8 m high, in dryish rain-forest or brigalow-bonewood scrub on deep red-brown loam up to 750 m.

Closely related to *E. agallocha*, but quite distinct in its mostly smaller, much more closely crenulate and more distinctly (though still obtusely) acuminate leaves, and in its entirely different ecology. It apparently never grows as a mangrove.

Excoecaria parvifolia *Muell. Arg.* in Flora 47: 433 (1864) & in DC.: 1221 (1866); Benth.: 153 (1873); Bailey: 1457 (1902); Pax & Hoffm. v: 165 (1912); Gardner: 72 (1931). Type: N.T., 'In Nova Hollandia septentrionali, in Arnhemsland', *F. Mueller* (MEL).

E. agallocha var. *muelleriana* Baill. in Adansonia 6: 325 (1866). Type: 'common in many places round the Gulf of Carpentaria and in Arnhemsland', 1855–6, *F. Mueller* (MEL).

WA (K); **NT** (DG, BT); **Q** (CK)—Endemic.

Shrub or tree of 4–6 m, in sandy alluvium on river bank or edge of lagoon up to 240 m.

The entire, cuneate-obovate, spathulate or narrowly oblong (strap-shaped) leaves of this plant, borne in sessile tufts or short brachyblasts along the rather stout and rigid branches, are unmistakable.

Fontainea *Heckel*

Rather closely related to *Dimorphocalyx*, differing principally in its non-accrescent calyx (but accrescence not invariable in *Dimorphocalyx*!), its densely white-pubescent or lanuginose petals, and its indehiscent, glabrous or pubescent, drupaceous fruit with a sharply 3–5(–6)-angled endocarp.

1. Midrib glabrous beneath; base of lamina gradually attenuate and decurrent on to the petiole; pubescence of inflorescence weaker **F. pancheri**
1. Midrib adpressed-pilose beneath; base of lamina ± abruptly contracted to the petiole; pubescence of inflorescence stronger . **F. picrosperma**

Fontainea pancheri (*Baill.*) *Heckel*, Étude sur le Fontainea (Thèse Fac. Méd. Montpellier): 11(?) (1870); Baill. in Adansonia 11: 80 (1873), *in obs.*, & Hist. Pl. 5: 194 (1874); Pax in Engl. & Prantl, Nat. Pflanzenf. III. 5: 87 (1890); Bailey: 1440 (1902); Pax in Engl., Pflanzenr. IV. iii: 30 (1911); Pax & Hoffm. in Engl. & Harms, Pflanzenf. ed. 2, 19c: 158 (1931); Guillaumin in Journ. Arn. Arb. 13: 92 (1932) & Fl. Anal. Nouv.-Caléd., Phan.: 188 (1948); Stern in Amer. Journ. Bot. 54: 668 (1967); Francis: 45, *in clavi*, 234 (1970); Airy Shaw in K.B. 29: 326 (1974). Syntypes: New Caledonia, *Pancher* 722; *Vieillard* 203, 204, 7; *Deplanche* 487 (all P).

Baloghia pancheri Baill. in Adansonia 2: 214 (1862); Benth.: 149 (1873).
B. drupacea Panch. ex Baill., *l.c.*: 215 (1862), *in obs.*, *nomen.*
Codiaeum ?? *pancheri* (Baill.) Muell. Arg.: 1117 (1866); Baill. in Adansonia 11: 80 (1873), *in obs.*

Q (WB, MO)—New Guinea, New Hebrides, New Caledonia.

Tree to 18 m at low altitudes (below 100 m; up to 1950 m in New Guinea).
Differs from *F. picrosperma* in its generally much weaker pubescence, and especially in the completely glabrous midrib. The base of the leaf is usually strongly attenuate and decurrent on to the petiole.
According to Baillon (1862), quoting field-notes of Pancher, the flowers are extremely fragrant, and the fleshy mesocarp of the fruit has a pleasant odour resembling that of plums or apricots, but the taste is exceedingly bitter.
Fontainea pancheri is confined in Queensland to a few localities in the south-east. I am treating for the present the Australian, New Guinea and New Hebrides populations as conspecific with the New Caledonian, but they show small differences among themselves and could perhaps deserve varietal recognition. (Cf. notes in K.B. 29: 327 (1974).)

Fontainea picrosperma *C. T. White* in Contrib. Arn. Arb. 4: 55 (1933); Hyland: 35 (1971). Type: Q., Boonjie, 1929, *Kajewski* 1262 (A).

Q (CK, NK)—Endemic.

Tree to 25 m high, common in rain-forest on red basaltic soil at 700–1050 m.
Dioecious, with a watery reddish or brownish or orange bark exudate. Leaves obovate to oblong-elliptic, 6–17 × 2·5–6 cm, cuneate to rounded and shortly trinerved at the base, abruptly and shortly caudate-acuminate at the

apex, cauda mostly acute, chartaceous, drying green, nervation prominent, especially beneath, quite glabrous above, densely or thinly adpressed-pilose on the nerves beneath; petiole slender, up to 4 cm long; stipules obsolete. Inflorescences mostly terminal, cymose, branched from the base, densely sericeous, the males many-flowered, up to 5(–6·5) cm long, the females few-flowered; pedicels 5–8 mm long. Calyx cupular, 2–3 mm long, very shortly 5-toothed. Petals broadly oblong-obovate, 5–6 × 2–3 mm, rounded at the apex, sericeous on both surfaces. Disk-glands reniform or semicircular. Stamens ± 30, inserted on a densely pilose convex receptacle; anthers oblong. Ovary subglobose, 2 mm diam., densely pilose, 5-locular; styles 5, deeply bifid, spreading at the base, then erect, 2 mm long. Fruit a hard, ellipsoid, shallowly 5-lobed drupe, 2–2·5 × 1·7 cm, softly adpressed-pilose, pinkish or apricot-tinged when ripe.

I have given a rather detailed description of this species, as the plant is not mentioned in Francis's *Australian Rain-Forest Trees* (eds 1951 & 1970) and is only briefly referred to by Hyland (*l.c.*, 1971). The more strongly sericeous inflorescences, and especially the adpressed-pilose midrib on the lower surface, are the most reliable differences from *F. pancheri*. Also the base of the leaf is more abruptly contracted or at least less gradually attenuate into the petiole. The two species are, however, closely related. *F. picrosperma* appears to be confined to the Atherton Tableland.

Glochidion *J. R. & G. Forst.*

Closely related to *Phyllanthus*, but differing in the total absence of disk-glands in the flower (both male and female), in the anthers mostly united into an oblong or ellipsoid mass, and in the styles united into a short or long cone or column, lobed or toothed at the apex, or into a clavate or globose structure, or very occasionally free; capsule sometimes large, multilocular, woody.

1. Inflorescence distinctly pedunculate and/or supra-axillary:
  2. Plant glabrous or almost so . **G. perakense** var. **supra-axillare**
  2. Plant shortly greyish tomentellous . . . . . . **G. philippicum**
1. Inflorescence epedunculate and strictly axillary:
  3. Style narrowly columnar, much exserted from calyx at time of flowering:
    4. Plant pubescent (see also *G. pruinosum* var.):
      5. ♀ flower of average size; style robust . . . . . **G. hylandii**
      5. ♀ flower very small; style slender . **G. ferdinandii** var. **pubens**
    4. Plant glabrous:
      6. Drip-tip not very sharp, often absent:
        7. ♀ flower very small, pedicel and style short and very slender **G. ferdinandii**
        7. ♀ flower of normal size, pedicel and style longer and thicker **G. sessiliflorum** var. **stylosum**
      6. Drip-tip very sharp; ♀ flower conspicuous, the style rather robust; lamina often ± conduplicate:
        8. Leaves conspicuously glauco-pruinose beneath, dull above; style 2 mm long, with divaricate lobes . . . . . **G. pruinosum**
        8. Leaves green, shining above; style 4 mm long, with erect lobes **G. pungens**

3. Style shortly conical or broadly button-shaped:
 9. Female flowers sessile or almost so, closely pulvinate (but capsules pedicelled):
 10. Leaf-undersurface with a smooth, 'thick', glabrous appearance, the minor nerves mostly immersed and inconspicuous **G. xerocarpum**
 10. Leaf-undersurface with a more distinctly 'veiny' appearance, with evident prominulous nervation:
 11. Leaves cuneate at base and decurrent on the petiole, glabrous; ♀ sepals glabrous . . . . **G. sessiliflorum**
 11. Leaves mostly broadly rounded at base, occasionally ± cuneate; ♀ sepals pilose:
 12. Plant glabrous or almost so . . . . **G. disparipes**
 12. Plant shortly pubescent . . . . **G. apodogynum**
 9. Female flowers evidently pedicellate:
 13. Styles forming a relatively large, prominent, ± hemispherical button on top of the ovary:
 14. Plant usually very shortly greyish-tomentellous; leaves often very asymmetrical at base; capsule deeply depressed, divided into numerous narrow segments . . . . **G. philippicum**
 14. Plant practically glabrous, very smooth; leaves symmetrically cuneate at base; capsule much less depressed, 6–8-locular **G. sessiliflorum** var. **pedicellatum**
 13. Styles short, free or connivent, not button-shaped, sometimes scarcely prominent:
 15. Leaves 7–17 × 3–7 cm, not glaucous beneath; plant glabrous except in *G. harveyanum* var. *pubescens*:
 16. ♀ pedicels ± stout, straight, ± erect, angled; ♀ sepals ± broad, 2–3·5 mm long, subacute, erect . . . . . **G. harveyanum**
 16. ♀ pedicels more slender; ♀ sepals small, 1·5 mm long, subobtuse **G. barronense**
 15. Leaves 3–9 × 1·5–3·5 cm, sometimes glaucous beneath; ♀ pedicels slender, often spreading, not noticeably angled; ♀ sepals smaller, ± obtuse, spreading; stems minutely puberulous:
 17. Leaves manifestly glaucous beneath; capsule deeply divided into rounded, almost separate cocci . . . . . **G. lobocarpum**
 17. Leaves slightly glaucous beneath, usually drying reddish-brown above; capsule with shallowly divided lobes **G. benthamianum**

Glochidion apodogynum *Airy Shaw* in K.B. 27: 44 (1972) & in Muelleria 4: 208 (1980). Type: Papua, *Brass* 24333 (K).

Q (CK, SK)—New Guinea (Papua).

Shrub or small tree of 2–4 m, among granite boulders in open forest at 560–800 m.

Closely related to *G. disparipes*, and perhaps not specifically distinct, but differing in its much stronger short pubescence.

Known from near Almaden, southeast of Chillagoe, on a granite hill, from the Cauley State Forest, between Mackay and Proserpine, on granite-derived soil, and from near Clairview, on Broad Sound, lat. 22° 10′ S.

GLOCHIDION ADAMII (*Muell. Arg.*) *C. A. Gardn. ex Beard*, W. Austr. Pl.: 58 (1967?) = **Sauropus glaucus** (*F. Muell.*) *Airy Shaw*: vide p. 675, infra.

Glochidion barronense *Airy Shaw* in K.B. 31: 343 (1976) & in Muelleria 4: 209 (1980). Type: Q., 'S.F.R. 191 Barron', *Irvine* 103 (K).

Q (CK)—Endemic.

Shrub or small tree of 1–6 m, in rain-forest at 750–800 m.

Closely related to *G. harveyanum*, and again perhaps not deserving specific rank, but the small ♀ sepals, only 1·5 mm long, obtuse at the apex, and leaving the upper part of the ovary exposed, give the impression of a distinct taxon.

So far only known definitely from SFR 191 Barron on the Atherton Tableland.

Glochidion benthamianum *Domin*: 872 [318] (1927); Airy Shaw in K.B. 31: 343 (1976); Airy Shaw in Muelleria 4: 209 (1980). Type: Q., Rockingham Bay, *Dallachy s.n.* (isotype K).

Phyllanthus ferdinandi var. *mollis*, subvar., Benth.: 97 (1873).
Glochidion ferdinandi var. *mollis*, subvar., Bailey: 1421 (1902).
G. capitis-york Airy Shaw in K.B. 23: 25 (1969) & 27: 68, *in adnot.*, 70, *in clavi et enum.* (1972). Type: Q., Iron Range, 1948, *Brass* 19254 (holotype K).
[*G. eucleoides* sec. Airy Shaw in K.B. 27: 59 (1972), quoad synon. et specim. Austral., *non* S. Moore.]

Q (CK, NK)—SE West New Guinea, Papua.

Shrub or small tree of 2–5 m, abundant in open savanna-forest adjacent to rain-forest, on thinly timbered river-banks, on edge of lagoons, in riverine rain-forest, or in red stony soil on hill, at 20–200 m; perhaps favoured by fire.

A slender, much branched, small-leaved species; branchlets minutely puberulous; leaves not exceeding 6 × 2·5 cm, glabrous or the midrib minutely puberulous, smooth and shining and reddish-brown above when dry, often slightly glaucous beneath, chartaceous, usually very unequal at the base, obtusely cuspidate or shortly acuminate at the apex. Male and female pedicels minutely puberulous, 3–5 mm long or rarely more. Capsule much depressed, 7–10 mm diam., 4–5 mm deep, much impressed at base and apex, minutely puberulous, shallowly 6-coccous, each coccus with a shallow dorsal groove; styles 0·5 mm long.

Differs from the somewhat similar *G. lobocarpum* in the asymmetrical base of the leaves, their much less glaucous undersurface and their reddish-brown colour above when dry, and in the very shallowly lobed capsules. The plant barely extends into North Kennedy District, south of Atherton.

GLOCHIDION CRASSIFOLIUM (Muell. Arg.) C. A. Gardn. ex Beard, l.c. supra = **Sauropus crassifolius** (*Muell. Arg.*) *Airy Shaw* (vide p. 672, *infra*).

Glochidion disparipes *Airy Shaw* in K.B. 27: 43, 65 (1972) & 31: 345 (1976) & Muelleria 4: 209 (1980); George & Kenneally in Miles & Burbidge, Biol. Surv. Prince Regent River Reserve: 47 (1975) & in Kabay & Burbidge,

Biol. Surv. Drysd. River Nat. Park: 54 (1977). Type: Papua, Cape Vogel Peninsula, *Hoogland* 4705 (holotype K).

[*G. ferdinandii* sec. Specht 3: 252, 398, 461 (1958), pro parte, *non* (Muell. Arg.) F. M. Bailey.]

WA (K); **NT** (DG); **Q** (CK, NK, PC)—New Guinea (NE and Papua).

Shrub or small tree of 4–7 m, in mixed open forest near sandstone escarpment, on edge of palm forest near hot spring, in rain-forest remnant, in strand-forest, on river levee, or in grassland, at low altitudes.

The round-based, broadly ovate-oblong leaves and small dense pulvini of sessile female flowers are the distinctive marks of this species. It is distinguished further from *G. apodogynum* by its almost glabrous condition, and from *G. xerocarpum* by the thinner texture and more veiny surface of the leaves. The species has been found as far south as Callide, some 260 km to the south of Rockhampton.

GLOCHIDION ECRASSIFOLIUM C. A. Gardn. ex Beard, sphalm. = *G. crassifolium* (Muell. Arg.) C. A. Gardn. ex Beard = **Sauropus crassifolius** (*Muell. Arg.*) *Airy Shaw*; *vide* p. 672, *infra*.

Glochidion ferdinandii (*Muell. Arg.*) *F. M. Bailey*: 1423 (1902); C. A. Gardn. ex Beard, W. Austr. Pl.: 58 (?1967); Anderson: 216 (1968); Francis: 219 (1970), *p.p.*; Beadle, Evans & Carolin: 251 (1972); Airy Shaw in K.B. 31: 346 (1976) & in Muelleria 4: 209 (1980). Type: N.S.W., Hastings River, *F. Mueller* (MEL).

Phyllanthus ferdinandi Muell. Arg. in Flora: 379 (1865) & in DC.: 300 (1866); Benth.: 96 (1873) (excl. vars. *supra-axillaris* & *mollis*); Baker, Hardw. Austr.: 355 (1919); Maiden, For. Fl. NSW 8 (73): 44, t. 278 (1923).

var. **ferdinandii**

P. ferdinandi var. *minor* Benth., *l.c.* (1873). Syntypes: N.S.W., Bremer River, *A. Cunningham* (MEL, K); New England, *C. Stuart* (MEL).

Q (SK, LT, PC, WB, MO); **NSW** (NC, CC, SC)—Endemic.

Glabrous shrub or tree to 8 m, in dense rain-forest, gallery forest, dry sclerophyll forest, sandstone gorge or brush gully, partly cleared woodland, moist shady spots, up to 880 m. Fig. 3D.

The diagnostic feature of *G. ferdinandii* is the very small axillary female flower on a short slender pedicel, with a much exserted short slender style. The leaves are oblong-elliptic, almost equally narrowed at base and apex, chartaceous to coriaceous, smooth and somewhat shining above, sometimes asymmetric at base. The species is now known as for north as Cawley State Forest, W of Cathu, lat. 20° 48′ S.

var. **pubens** *Maiden ex Airy Shaw* in K.B. 31: 346 (1976). Type: N.S.W., Hawkesbury River, 1888, *Betche* in NSW 128619 (NSW).

NSW (CC)—Endemic.

Brush tree of 4–6 m, in rain-forest.

Differs from the typical glabrous form in the short pubescence of all parts.

Glochidion harveyanum *Domin*: 873 [319] (1927). Type: Q., Regenwälder bei Harveys Creek, 1910, *Domin* s.n. (PR).

var. **harveyanum.**

Q (CK, NK)—Endemic.

Glabrous shrub or tree 2–12 m tall, in rain-forest on red soil, forest margins, edge of lake, creek bank in open forest, or dry scrub, at 15–1750 m.

The rather stout, straight, erect, angled pedicels of the female flower, with relatively broad, subacute, suberect sepals, distinguish *G. harveyanum* from all others.

var. **pubescens** *Airy Shaw* in K.B. 31: 347 (1976). Type: Q., Davies Creek, *Hyland* 7722 (K).

Q (CK)—Endemic.

Tree of 10 m, in gallery forest at 420 m.

Pubescent throughout; superficially similar to *G. hylandii*, but distinguished by its short included style, as well as by the pedicels and sepals.

Glochidion hylandii *Airy Shaw* in K.B. 31: 347 (1976) (excl. *Hyland* 7799 & 7932) & in Muelleria 4: 209 (1980). Type: Q., Forest Reserve 185, 1961, *Hyland* 02224 (holotype K).

Q (CK)—Endemic.

Many-stemmed shrub or small tree of 7–10 m, in rain-forest at 700–1050 m.

Related to *G. harveyanum*, and especially its var. *pubescens*, differing in its slender, scarcely angled ♀ pedicels, much smaller ♀ sepals, conspicuously exserted puberulous style up to 3 mm long, and larger somewhat inflated-looking capsules. The leaves are sometimes distinctly glaucescent beneath.

Apparently confined to the Atherton Tableland. The two collections excluded above (*Hyland* 7799 & 7932) are referable to *G. apodogynum* Airy Shaw, *q.v.*

Glochidion lobocarpum (*Benth.*) *F. M. Bailey*: 1424 (1902) & Compreh. Cat. Queensl. Pl.: 476 (1913); Airy Shaw in K.B. 27: 41, 72 (1972) & 31: 348 (1976) & in Muelleria 4: 209 (1980).

Phyllanthus lobocarpus Benth.: 97 (1873); F. M. Bailey, Syn. Queensl. Fl. 465 (1883). Syntypes: Q., Rockhampton, *O'Shanesy* (MEL); Nerkool Creek (± 20 km W of Rockhampton), *Bowman* (K, MEL).

Q (SK, PC)—E New Guinea.

Shrub or small tree to 7 m tall, in open or beach forest up to 30 m.

A very distinct species from its glabrous, narrowly elliptic-oblong leaves, glaucous beneath, and the deeply lobed capsule, almost completely divided into 6 obovoid or subglobose cocci. So far known only from the area between Townsville and Rockhampton.

Glochidion perakense *Hook. f.* var. **supra-axillare** (*Benth.*) *Airy Shaw* in K.B. 27: 72 (1972) & 31: 349 (1976) & in Muelleria 4: 210 (1980). Syntypes: Q., Rockingham Bay, *Dallachy* (MEL, K); Rockhampton, *Thozet*, *O'Shanesy* (MEL).

Phyllanthus ferdinandi var.? *supra-axillaris* Benth.: 96 (1873); Maiden & Betche in Proc. Linn. Soc. NSW ser. 2, 8: 315 (1894).
Glochidion ferdinandi var. *supra-axillaris* (Benth.) Bailey: 1424 (1902).
G. supra-axillare (Benth.) Domin: 872 [318] (1927).

NT (DG); **Q** (CK, NK, SK, MO)—New Guinea, Solomon Is.

Shrub or tree to 15 m high, in rain-forest or seasonal swamp or open *Eucalyptus* forest up to 1050 m.

The pedunculate and/or supra-axillary inflorescences usually distinguish this species from all others in Australia. From *G. philippicum*, in which the inflorescences occasionally show this character, *G. perakense* differs in its 'normal', 5–6-locular capsule. It is also most commonly glabrous, but there is some variation in this feature. An unusually strongly pubescent form from the Northern Territory (Coomalie Creek, *Must* 1352) approaches *G. angulatum* C. B. Rob. (Philippines, Sabah, Celebes, Moluccas and New Guinea), and requires further study from fruiting material.

Glochidion philippicum (*Cav.*) *C. B. Rob.* in Philipp. Journ. Sci. 4, Bot.: 103 (1909); Airy Shaw in K.B. 27: 58, 73 (1972) & 31: 349 (1976) & in Muelleria 4: 210 (1980). Type: Philippine Is., inter Oaz et Irraga, *Ludovicus Née* (MA?).

Bradleia philippica Cav., Ic. 4: 48, t. 371 (1798).
Phyllanthus ferdinandi var.? *mollis* Benth.: 97 (1873). Type: Q., Rockingham Bay, *Dallachy* (MEL, K).

NT (DG); **Q** (CK, NK)—Formosa, Philippines, Talaud Is., Java and throughout E Malesia to the Solomons.

Shrub or small tree to 15 m tall, in rain-forest or fringing forest or monsoon forest, on sandy loam or loam from granite, up to 660 m.

An unmistakable species when in fruit, from the numerous (10–12) narrow segments into which the flattened capsule is divided. The inflorescences are sometimes peduncled and/or supra-axillary; the styles form a conspicuous button on top of the ovary; the leaf-base is asymmetrical, and varies from cuneate to broadly rounded. The whole plant is generally shortly greyish-tomentellous.

Glochidion pruinosum *Airy Shaw* in Muelleria 4: 211 (1980). Type: Q., Thornton Peak, 1973, *Hyland* 7064 (holotype K).

Q (CK)—Endemic.

Shrub or small tree to 10·5 m, in montane rain-forest bordering on heath at 1260–1560 m.

A species of high altitudes, more or less intermediate between *G. ferdinandii* and *G. pungens*. The female flower and exserted style are midway in size

FIG. 3. *Glochidion pungens:* **A1** habit × ⅔, from *Irvine* 1693; **A2** female flower × 4, from *Irvine* 1693. *Homalanthus novoguineensis:* **B1** habit × ⅔, from *Brass* 2006; **B2** fruit, from *Brass* 2006. *Homalanthus stillingiifolius:* **C1** habit × ⅔ from *Coveny* 9376 & *Hubbard* 3732; **C2** fruit × 2, from *Coveny* 9376. *Glochidion ferdinandii:* **D1** habit × ⅔, from *Hubbard* 5412; **D2** flower × 6, from *Hubbard* 4695; **D3** fruit × 1, from *Hubbard* 5412.

between the two. The leaf-apex is sharply pointed, almost as in *G. pungens*, but in the pruinose undersurface it differs from both species. There is probably some relationship between these taxa and *G. merrillii* C. B. Rob. (Philippines) and *G. alticola* Airy Shaw (Sumatra).

Glochidion pruinosum is so far known from only two localities—Thornton Peak (type collection) and Mt Bellenden-Ker. In the only known collection from the latter station there is a development of puberulence on the branchlets and female pedicels.

Glochidion pungens *Airy Shaw* in K.B. 31: 349 (1976). Type: Q., Mt Lewis Range, 1964, *Schodde* 4152 (holotype BRI).

Q (CK)—Endemic.

Tree of unknown stature (trunk 70 cm dbh), in rain-forest at 1200 m. Fig. 3A.

The conspicuously exserted, 4 mm long style and pungent drip-tip of the leaves mark off this species from all others in Australia, except *G. pruinosum*, which is more or less intermediate between the present plant and *G. ferdinandii* and differs further in the pruinose undersurface of the leaves. These montane species appear to be relicts, persisting in a few high-altitude localities.

GLOCHIDION RHYTIDOSPERMUM (F. Muell. ex Muell. Arg.) Hj. Eichl., Suppl. Black's Fl. S. Austr.: 210 (1965); F. Muell. ex Beard, W. Austr. Pl.: 58 (?1967) = **Sauropus trachyspermus** (*F. Muell.*) *Airy Shaw*; vide p. 685 *infra*.

GLOCHIDION RIGENS (F. Muell.) Hj. Eichl., l.c. (1965) = **Sauropus rigens** (*F. Muell.*) *Airy Shaw*, vide p. 683, *infra*.

Glochidion sessiliflorum *Airy Shaw* in K.B. 31: 350 (1976) & in Muelleria 4: 211 (1980). Type: Q., Claudie River, *Hyland* 7805 (holotype K).

var. **sessiliflorum:** floribus femineis sessilibus, ovario glabro, stylo brevissimo pulviniformi.

QUEENSLAND. Cook District (Cape York Peninsula): Claudie River, *Hyland* 7805 (type); Timber Reserve 14 (Leo Creek Road), *Hyland* 2658.

var. **pedicellatum** *Airy Shaw* var. nov., floribus femineis usque 5 mm pedicellatis, ovario puberulo, stylo brevissimo pulviniformi.

QUEENSLAND. Cook District (Atherton Tableland): S.F.R. 251, Charmillin Logging Area, 17° 40′ S, 145° 30′ E, clearing in rain-forest, 750 m, 29 Sept. 1976, *Dockrill* 1271 (holotype K):—Shrub 3 m tall; flowers cream; fruit greenish yellow, some with pink suffusions. Rooty L. A., *Hyland* 3392 & 3393 RFK (see Muelleria, *l.c. supra*). S.F.R. 185, Robson L. A., 17° 07′ S, 145° 37′ E, rain-forest, 740 m, 21 Oct. 1976, *Dockrill* 1306:—Small tree 5 m; flowers cream. S.F.R. 756, Lower Koolmoon L. A., 17° 45′ S, 145° 34′ E, rain-forest, 720 m, 4 Nov. 1976, *Hyland* 9182:—Small multistemmed tree with creamy yellow male flowers.—North Kennedy Distr.: Sea View Range, *Dallachy* 23 & 31 (MEL) (see Muelleria, *l.c. supra*.).

var. **stylosum** *Airy Shaw* var. nov., floribus femineis brevissime pedicellatis, ovario puberulo, stylo breviter crasse columnari 1–1·5 mm longo.

QUEENSLAND. Cook District (Atherton Tableland): T.R. 146, Tableland L. A., 15° 45′ S, 145° 15′ E, rain-forest, 660 m, 9 July 1975, *Hyland* 3221 RFK (holotype K):—Tree; trunk 50 cm d.b.h., buttressed, fluted; bark tessellated, flaky.

It is not impossible that further material of vars. *sessiliflorum* and *stylosum* may show that the above 3 taxa deserve specific rank, although vegetatively they seem indistinguishable. The foliage of all three is quite glabrous, and when dry assumes a smooth appearance and a rather deep green colour, which can best be described as 'oily'.

GLOCHIDION THESIOIDES Hj. Eichl., l.c. (1965) = **Sauropus thesioides** (*Hj. Eichl.*) *Airy Shaw*; vide p. 684 *infra*.

GLOCHIDION TRACHYSPERMUM (F. Muell.) Hj. Eichl., l.c. (1965) = **Sauropus trachyspermus** (*F. Muell.*) *Airy Shaw*; vide p. 685, *infra*.

Glochidion xerocarpum (*O. Schwarz*) *Airy Shaw* in Muelleria 4: 212 (1980; Euph. New Guinea: 115 (1980). Type: N.T., Darwin, Mindel Beach, 1927, *Bleeser* 495 (B†; isotype NSW).

Phyllanthus xerocarpus O. Schwarz in Fedde, Rep. Sp. Nov. 24: 87 (1927).
[*G. ferdinandii* sec. Specht: 252, 398, 461 (1958), *pro parte, non* (Muell. Arg.) F. M. Bailey.]
[*G. mindorense* subsp. *mindorense* sec. Airy Shaw in K.B. 27: 21 (non 66 nec 72) (1972) & 29: 291 (1974), *pro majore parte*, non *G. mindorense* C. B. Rob.]
[*G. disparipes* sec. Airy Shaw in K.B. 31: 345 (1976), quoad *Specht* 24 & 860, *non* Airy Shaw *sens. strict.* (1972).]

NT (DG)—Lesser Sunda Is., Java, Moluccas, Celebes, Sabah, S Philippines, New Guinea.

Small slender tree to 10 m high, in rain-forest, or in monsoon forest on truncated lateritic podsol, or in deciduous vine thicket on stabilized coastal sand-dunes, at very low altitudes.

Closely related to *G. disparipes*, but differing in the much less raised, often almost immersed minor venation, giving the leaves a thickish, smooth appearance, especially on the underside. Throughout its range the plant seems always to grow close to or not far from the sea.

Homalanthus *Juss.*

Shrubs or trees (sometimes exceeding 20 m), mostly glabrous. Leaves broadly or narrowly ovate, occasionally peltate, petiolate, mostly with 2 prominent, distinct or confluent, superior or inferior glands at the apex of the petiole; stipules lanceolate, brown-membranous, conspicuous. Flowers monoecious; inflorescences racemose, male or female or mixed, bracts usually 1–2-glandular at the base. Male flower; calyx 1–2-valved (mostly

2-valved in Australia), compressed; stamens 5–50, filaments short, disk 0, pistillode 0. Female flower; calyx disciform or cupuliform, shortly 2–3-lobed; ovary bilocular or trilocular; styles 2–3, connate below, longer and divaricate or short and recurved, entire or shortly bifid, sometimes bearing a gland on the lower side at the apex. Fruit capsular or indehiscent, 2–3-locular; seeds ovoid, carunculate.

1. Slender shrub; leaves deltoid, up to 3·5 × 3 cm, papillose-pilosulous beneath; inflorescence very delicate; floral bracts usually without dorsal glands; capsule often bearing a pair of short dorsal wings on each loculus **H. stillingiifolius**
1. More robust shrubs or trees; leaves triangular-ovate, 5–15 cm long and broad, glabrous beneath; floral bracts with dorsal glands; capsule without dorsal wings:
   2. Bract-glands compound, 'tremelloid', foveolate; style-branches very short and strongly recurved . . . . . . **H. novo-guineensis**
   2. Bract-glands simple, not 'tremelloid'; style-branches elongate, erect or spreading . . . . **H. populifolius**

Homalanthus novo-guineensis (*Warb.*) *Lauterb. & K. Schum.*, Fl. Deutsch. Schutzgeb. Südsee: 407 (1901); J. J. Sm. in Nova Guinea 8: 241 (1910); Pax & Hoffm. v: 48 (1912); Airy Shaw in K.B. 21: 410 (1968) (excl. specim. Buderim Mt, *Longman s.n.*) & in Muelleria 4: 238 (1980), *q.v.* Type: NE New Guinea, Finschhafen, 'am Sattelberg', 1889, *Hellwig* 555 (B†).

Carumbium novo-guineense Warb. in Engl., Bot. Jahrb. 18: 199 (1893).
Homalanthus populifolius sec. Hyland: 33 (1971), *p.p.*; George & Kenneally in Miles & Burbidge, Biol. Surv. Prince Regent River Reserve: 47 (1975); *non* Grah.

WA (K); **NT** (DG); **Q** (CK)—Moluccas, Lesser Sunda Is., New Guinea, Solomon Is.

Shrub or slender tree to 20 m high, in rain-forest on humic sand or in cadjeput forest in creek delta on silt, up to 900 m. Fig. 3B.

Closely related to *H. populifolius*, but distinguished by the very short and strongly recurved style-branches, and by the compound 'tremelloid' bract-glands. The areas of the two species approach, but apparently do not quite meet, near the boundary of Cook and North Kennedy districts in North Queensland. Further collections are needed from West Australia and the Northern Territory.

Homalanthus populifolius *Grah.* in Edinb. New Philos. Journ. [3]: 175 (1827); Hook. in Curt. Bot. Mag. 54: t. 2780 (1827); F. Muell., Fragm. 1: 22 (1858); Baill., Ét. Gén. Euphorb., Atlas: t. 8, figs. 22–26 (1858); Moore & Betche: 79 (1893); Bailey: 1454 (1902); Pax & Hoffm. v: 46 (1912); Anderson: 219 (1968); Francis: 234 (1970); Hyland: 33 (1971), *p.p.*; Beadle, Evans & Carolin: 253 (1972). Type: New Holland, 1824, *Fraser*, cult. Roy. Bot. Gard. Edinb., 1827, *Graham*.

Carumbium pallidum Muell. Arg. in Linnaea 32: 85 (1863). Type: 'In horto Berolinensi (?) colebatur', *leg.*? (B†).
C. platyneurum Muell. Arg., *ll. cc.* Type: N.S.W., 'In Nova Hollandia prope Sidney', *Vieillard* (P).
C. sieberi Muell. Arg. in Linnaea 32: 85 (1863) & in DC.: 1145 (1866); Baill. in Adansonia 6: 326 (1866). Type: 'In Nova Hollandia', *Sieber* 640 (whereabouts unknown).
[*C. populneum* sec. Muell. Arg.: 1144 (1866), p.p., *non* (Geisel.) Pax]
C. populifolium (Grah.) Benth.: 150 (1873); *non* Reinw. 1823.

Q (NK, PC, WB, MO); **NSW** (NC, CC, SC); **V** (E)—Papua (incl. Louisiades & Woodlark I.); Ceylon (? planted).

Shrub or tree to 7·5 m high, common in scrub and open rain-forest on red basaltic soil or sandstone or brown loam, often in gullies, up to 600 m.

Distinguished from *H. novo-guineensis* by the much longer, usually erect or spreading style-branches, and by the small, simple, often inconspicuous bract-glands. *H. populifolius* extends southwards from the southern edge of the area of *H. novo-guineensis* into southeast New South Wales and the extreme east of Victoria. It occurs also in southeastern New Guinea and certain of the nearer island groups to the east.

Homalanthus stillingiifolius *F. Muell.*, Fragm. 1: 32 (1858); Moore & Betche: 79 (1893); Bailey: 1454 (1902); Pax & Hoffm. v: 53 (1912); Anderson: 219 (1968); Francis: 234 (1970). Type: Q., 'Ad flumen Brisbane', *Hill & Mueller* (MEL).

Wartmannia stillingiaefolia (F. Muell.) Muell. Arg. in Linnaea 34: 219 (1865) & in DC.: 1147 (1866).
Carumbium stillingiaefolium (F. Muell.) Baill. in Adansonia 6: 325 (1866); Benth.: 150 (1873).

Q (MO); **NSW** (NC, CC, CT)—Endemic.

Slender shrub to 3 m high, on edge of semi-rain-forest and in Eucalyptus forest on stony mountain slopes and on rocky hillsides or among large sandstone boulders below tall cliffs, up to 750 m. Fig. 3C.

Stems and inflorescences far more slender and delicate, and leaves smaller and more deltoid, than in the other two species, the leaves shortly white-papillose-pilosulous beneath. Bracts without or occasionally with a very small dorsal gland; ♂ flowers numerous, very small; sepal 1, or sometimes 2 in the central flower of each triad; stamens 3–7; ♀ flowers 0–3 at the base of each inflorescence, on long slender pedicels; ovary bilocular, with rather long simple styles; capsule 5 mm diam., each loculus commonly bearing a pair of shortly triangular ± ascending dorsal wings or processes (this latter feature is apparently unique in the genus).

Hylandia *Airy Shaw*

Tall tree, with obovate, elliptic or ± oblong, stiffly coriaceous, rather long-petioled leaves. Flowers dioecious or occasionally monoecious, with conspicuous petals, borne in many-flowered (male) or few-flowered (female) pyramidal thyrses. Stamens 10–16, biseriate, shortly connate. Ovary and fruit bilocular, fruit 2 cm diam.

FIG. 4. *Hylandia dockrillii*. **A** habit × ⅔, from *Hyland* 6701; **B** young fruit & leaves × ⅔, from *Hyland* 5933; **C** flower × 2, from *Hyland* 6701; **D** young fruit × 2, from *Hyland* 5933; **E** fruit × ⅔, from *Smith* 10105.

Hylandia dockrillii *Airy Shaw* in K.B. 29: 329 (1974). Type: Q., S.F.R. 756, *Dockrill* 25 (holotype K).

Q (CK)—Endemic.

Tree to 30 m high, in rain-forest at 680–760 m. Fig. 4, & cover illustration of separate offprint.

Hylandia is related to *Dimorphocalyx*, and less closely to *Trigonostemon* and *Fontainea*, differing from all three in its tall arborescent habit, bilocular ovary and relatively large (2 cm diam.) bilocular fruit. The flowers are conspicuous from their white broadly spathulate petals 6–8 mm long, which are golden-sericeous externally and reddish-pilose within. In the male flower there are 5 subglobose disk-glands. In the female the ovary is densely long-fulvous-pilose, with 2 deeply bifid narrow or flabellate styles. Occasionally a few female flowers occur in male inflorescences. Mature dehiscing fruit and seeds have not yet been collected.

The tree has been known in the forestry service by the trade name of 'blushwood'.

Leptopus *Decne*

Monoecious woody annuals or short-lived perennials or subshrubs, with much of the aspect of *Phyllanthus*, from which they differ in the presence of petals and of a pistillode in the male flower. From *Actephila*, which also (usually) has petals, and a pistillode, *Leptopus* differs in its Phyllanthoid habit and in the stamens alternating with the petals, rather than arising near the centre of the broad disk; from *Securinega*, which also has a pistillode, it differs in the habit, in the presence of petals, and in the strictly capsular, not baccate, fruit. Flowers in axillary fascicles, the males minute, the females with broad sepals.

Leptopus decaisnei (*Benth.*) *Pojarkova* in Not. Syst. Herb. Inst. Bot. Acad. Sci. URSS 20: 271 (1960). Type: Timor, *Decaisne* (P).

[*Andrachne fruticosa* sec. Decne in Nouv. Arch. Mus. Paris 3: 484 (1834); Muell. Arg.: 235 (1866); Baill. in Adansonia 6: 334 (1866); Pax & Hoffm. xv: 173 (1922); Backer & Bakh. f.: 470 (1963); *non* L.]

A. decaisnei Benth.: 88 (1873); Bailey 1415 (1902); Ewart & Davies: 162 (1917); Gardner: 72 (1931).

WA (K); **NT** (DG, VD); **Q** (LT)—Java (Madura), Timor.

var. **decaisnei**

Erect herb of 50–60 cm, on heavy clay loam or sandy loam on river levee at low altitude.

An annual or short-lived perennial with a woody base (cf. *Sebastiania chamaelea* and *Sauropus bacciformis* (L.) Airy Shaw). Leaves cuneate-obovate, up to 4 × 2·5 cm, but those on lateral branches often small and orbicular, as in var. *orbicularis*.

var. **orbicularis** (*Benth.*) *Airy Shaw* in K.B. 32: 379 (1978). Type: W.A., Port Walcot, *C. Harper* (K).

Andrachne decaisnei var. *orbicularis* Benth., *l.c.* (1873).
A. fruticosa var. *orbicularis* (Benth.) Pax & Hoffm. xv: 173 (1922).
A. orbicularis (Benth.) Domin: 869 [315] (1927), *non* Roth (1821).
Arachne orbicularis (Benth.) Pojark. in Bot. Zhurn. SSSR 25: 342 (1940).

WA (N-Fort); **NT** (R); **Q** (LT)—Papua.

Herb of 20–25 cm, rare in low open forest on red sandy loam at low altitude.

All leaves small and suborbicular, 5–10 mm diam. Very doubtfully worth distinguishing from *L. decaisnei*, even as a variety, but few collections available.

Macaranga *Thou.*

Differs from *Mallotus* in the following characters: indumentum of simple or fascicled, never of truly stellate hairs, very rarely of minute scales; leaves never opposite, always ± glandular-granular beneath; stipules sometimes large and conspicuous; floral bracts often bearing large patellar glands; disk 0; stamens few, sometimes 1, occasionally up to 20; anthers 3–4-locellate; ovary 1–6-locular; style(s) short or elongate, when solitary almost gynobasic.

1. Leaves deeply peltate, very broadly ovate or suborbicular, palmately nerved, up to 30 cm diam. . . . . **M. tanarius**
1. Leaves not or only slightly peltate:
  2. Leaves palmately nerved, broadly ovate-rhomboid, up to 115 cm long; stipules linear-subulate . . . . **M. involucrata** var.
  2. Leaves pinnately nerved:
    3. Capsule muricate; styles filiform, 1 cm or more long:
      4. Plant almost glabrous; leaves densely (contiguously) minutely lepidote beneath (cf. *M. chlorolepis* Airy Shaw, New Guinea) **M. subdentata**
      4. Plant ± shortly tomentellous, especially branchlets and inflorescences; leaves rather distantly granular-glandular-punctate beneath **M. inamoena**
    3. Capsule smooth; styles 1–2 mm long:
      5. Petiole less than 2 cm long; ♀ inflorescence and infructescence branched only near apex, with ± long bare peduncle **M. dallachyana**
      5. Petiole up to 7 cm long; ♀ inflorescence and infructescence shortly branched to near base:
        6. Bracts of ♂ inflorescence 1–2 mm long, entire **M. inermis** (see under following sp.)
        6. Bracts of ♂ inflorescence 4–5 mm long, laciniate . **M. polyadenia**

Macaranga dallachyana (*Baill.*) *Airy Shaw* in K.B. 23: 90 (1969) & in Muelleria 4: 235 (1980). Type: Q., '*Dallachy* (1865), Rockingham's Bay, "salt water creeks" (herb. *F. Muell.*!)' (MEL).

Echinus dallachyanus Baill. in Adansonia 6: 314 (1866).
Mallotus dallachyi [sic] ('Baill.') F. Muell., Fragm. 6: 184 (1868). Type as above.
Macaranga dallachyi (Baill. ex F. Muell.) F. Muell. ex Benth.: 144 (1873); Bailey, Syn. Queensl. Fl.: 479 (1883) & Queensl. Fl. 5: 1450 (1902); Pax & Hoffm. vii: 394 (1914).
[*Mallotus polyadenus* sec. Pax & Hoffm. *l.c.*: 198 (1914), p.p., quoad synon. et specim. 'Rockingham Bay (Dallachy!)', non *M. polyadenos* F. Muell.]

Q (CK, NK)—Endemic.

Shrub or small tree in rain-forest or by salt water creeks; no further field notes.

The short petioles (less than 2 cm long) and the relatively long bare peduncle of the female inflorescence distinguish *M. dallachyana* from the other species with pinnately nerved leaves, short styles and smooth capsules. The leaves are minutely and densely but sometimes very obscurely, puncticulate-lepidote on the lower surface. The inflorescences and fruits have a minute ochraceous leprose-farinaceous covering. Apparently a very scarce species: besides the type-collection, cited above, I have only seen the following specimen, from the Atherton Tableland: Danbulla, Stony Creek logging area, 1957, *L. S. Smith* 10118.

Macaranga inamoena *F. Muell. ex Benth.*: 145 (1873); Bailey: 1451 (1902); Pax & Hoffm. xvii: 360 (1914); Hyland: 60 (1971); Airy Shaw in K.B. 31: 396 (1976), *in clavi*, & in Muelleria 4: 236 (1980). Type: Q., Rockingham Bay, *Dallachy* (K).

Q (CK, NK)—Endemic.

Small tree of 3–8 m, in rain-forest at 30–1100 m.

Distinguished from the very similar *M. subdentata* by the shortly tomentellous indumentum of the branchlets and inflorescences, and by the distantly granular-glandular undersurface of the leaves. As in *M. subdentata*, male and female inflorescences are sometimes produced from the apex of the same branchlet (e.g. *Hartley & Hyland* 14097). A rather common species on the Atherton Tableland.

Macaranga involucrata (*Roxb.*) *Baill.* var. **mallotoides** (*F. Muell.*) *Perry* in Journ. Arn. Arb. 34: 223 (1953); Airy Shaw in K.B. 31: 394 (1976). Type: Q., 'Ad montem Elliot', *Fitzalan* (MEL).

[*M. involucrata* sec. Baill., Ét. Gén. Euphorb.: 432 (1858) & in Adansonia 6: 317 (1866); Muell. Arg.: 1011 (1866); F. Muell., Fragm. 6: 183 (1868); Benth.: 146 (1873); Hyland: 60 (1971); vix *Urtica involucrata* Roxb.]
M. mallotoides F. Muell., Fragm. 4: 139 (1864); Pax & Hoffm. vii: 376 (1914).
M. asterolasia F. Muell., Fragm. 4: 140 (1864), *in obs.*; Baill. in Adansonia 6: 317 (1866). Type: Q., 'Ad montem Elliot', ?*Fitzalan* (MEL).

NT (DG); **Q** (CK, NK)—Endemic. (Var. *involucrata* in Moluccas, Tenimber Is., Kei Is., New Guinea.)

Shrub or tree of 3–18 m, in rain-forest (on sandy soil in NT) or gallery rain-forest at 10–120 m.

The palmately nerved leaves distinguish this species from all other Australian species except *M. tanarius*, from which it differs in the non-peltate or scarcely peltate insertion of the petiole, in the small inconspicuous floral bracts of the male inflorescence, and in the dicoccous ovary and capsule.

Macaranga polyadenia *Pax & Hoffm.* xiv: 25 (1919); Perry in Journ. Arn. Arb. 34: 249 (1953); Whitmore in Airy Shaw, K.B. Add. Ser. VIII: 154 (1980). Type: New Guinea, *Ledermann* 8664, 8677 (B†).

M. fimbriata S. Moore in Journ. Bot. Brit. & For. 61, Suppl.: 48 (1923); Perry in Journ. Arn. Arb. 34: 248 (1953); Airy Shaw in K.B. 31: 395 (1976) & in Muelleria 4: 236 (1980). Type: Papua, Sogere, 1885–6, *Forbes* 247 (BM).

M. multiflora C. T. White in Proc. Roy. Soc. Queensl. 55: 83 (1944). Type: Q., Johnstone River, *Rev. N. Michael* s.n. (BRI).

[*M. inermis* sec. Airy Shaw in K.B. 31: 395 (1976), *non* Pax & Hoffm.]

Q (CK, NK)—New Guinea, Solomon Is.

Tree of unknown stature (bole 20–40 cm d.b.h.), with prop-roots, very common in swamps, or in rain-forest, also on a red stony hill, up to 450 m.

Macaranga polyadenia and *M. inermis* Pax & Hoffm. are exceedingly difficult to distinguish vegetatively, though the leaves of *M. polyadenia* are often larger (especially broader) and longer-petioled than those of *M. inermis*. I now think that the specimens that I cited in 1976 as *M. inermis* are probably small-leaved examples of *M. polyadenia*. When in flower the two species are immediately recognizable by the relatively large fimbriate floral bracts of *M. polyadenia* and the small entire bracts of *M. inermis*. Also the fruiting pedicels of *M. polyadenia* are relatively short and stout, whilst those of *M. inermis* are longer and slenderer. The latter species is common in Papua, and there seems to be no reason why it should not occur also in north Queensland.

Macaranga subdentata *Benth.*: 145 (1873); Bailey: 1451 (1902); Pax & Hoffm. vii: 361 (1914); Hyland: 36 (1971); Airy Shaw in K.B. 31: 396 (1976) & Muelleria 4: 236 (1980). Type: Q., Rockingham Bay, *Dallachy* (K).

Q (CK, NK)—Endemic.

Small tree 6–10 m high, in mesophyll vine-forest on red clay soil or in complex mesophyll vine-forest in gully on soils derived from granite, at 200–230 m. Fig. 5D.

Differs from *M. inamoena* in being almost glabrous, and in the very densely and minutely stellate-lepidote undersurface of the leaves, the margins of which may be repand-dentate, obscurely sinuate or subentire. Inflorescences closely fascicled from distal axils, the males simply (pseudo-) spicate, floriferous over most of their length, up to 15 cm long, the females bare for at least the lower half, with a strongly flattened rhachis, and with a few ascending branches from their upper half, these floriferous only at their apex; male and female inflorescences sometimes arising from the same branch;

occasionally even bisexual inflorescences are produced (*L. S. Smith* 11121). Capsules densely minutely rufous-lepidote, with numerous conical warts; styles elongate-filiform, 'whiskery', up to 13 mm long.

Macaranga tanarius (*L.*) *Muell. Arg.*: 997 (1866); Benth.: 146 (1873); Moore & Betche: 79 (1893); Bailey: 1452 (1902); J. J. Sm.: 496 (1910); Pax & Hoffm. vii: 352 (1914); Merr., Interpr. Rumph. Herb. Amboin.: 320 (1917); Backer & Bakh. f.: 488 (1963); Anderson: 219 (1968); Francis: 230 (1970); Hyland: 71 (1971). Type: Amboina, *Tanarius minor* Rumph., Herb. Amboin. 3: 190 (1743).

Ricinus tanarius L. in Stickm., Herb. Amboin.: 14 (1754) & in Amoen. Acad. 4: 125 (1759).
Mappa tanarius (L.) Bl., Bijdr.: 624 (1825); ('*tanaria*') Baill. in Adansonia 6: 316 (1866).

NT (DG); **Q** (CK, NK, SK, LT, PC, WB, BT, MO); **NSW** (Far NC)—Andamans, Nicobars, SE & Lower Siam, Cochinchina, S China, Formosa & Ryu-Kyu Is., and throughout Malesia to Melanesia.

Shrub or tree to 8 m tall, gregarious locally on sandy river floodbanks, or in fringing forest, or on sea-beaches, or on sand cays or shingle-mangrove islands, up to 420 m.

The deeply peltate, orbicular-ovate leaves distinguish *M. tanarius* from all other Australian species. The conspicuously fringed floral bracts and the long spidery processes of the rather large trilocular capsules are also distinctive. The species is evidently frequent on islands of the Great Barrier Reef and other offshore islands of Queensland.

Bentham's locality, 'Liverpool River, *Gulliver*', listed by him under Queensland, should have been entered under N Australia. The Liverpool River enters the sea at Maningrida, on the north coast. *M. tanarius* seems to be scarce in the Northern Territory. Neither Gulliver's collection, nor Armstrong's from Port Essington, nor any more recent gathering from the Territory, is represented in the Kew Herbarium.

Mallotus *Lour.*

Shrubs or trees, rarely climbers, often with stellate hairs; leaves opposite (and then often unequal) or alternate, often with glandular granules on the lower (rarely also on the upper) surface, palminerved or penninerved, sometimes peltate, often with macular glands near the base; stipules mostly subulate. Inflorescences terminal or lateral, racemose or thyrsiform, unisexual, usually many-flowered, usually dioecious; flowers apetalous. Male flower: calyx closed in bud, valvately 3–4-partite at anthesis; stamens mostly numerous, filaments free, connective sometimes truncate and subpeltate; disk 0 or central and disciform or represented by discrete glands. Female flower: calyx ± 3–5-lobed, sometimes spathaceous and caducous; disk 0; ovary (2–)3(–4)-locular; styles simple, plumose. Capsule (2–)3(–4)-locular, smooth or variously echinate, dehiscent or rarely indehiscent; seeds mostly smooth.

1. Leaves with glandular granules above as well as below; leaves variably alternate or opposite; cocci of capsule smooth, rounded and deeply separated (§ *Polyadenii*) . . . . **M. polyadenos**
1. Leaves without glandular granules on upper surface:
 2. ♀ calyx spathaceous; leaves variably alternate or opposite; dried plant smelling of fenugreek (§ *Stylanthus*) . . . . **M. oblongifolius**
 2. ♀ calyx not spathaceous; leaves either strictly alternate or strictly opposite; dried plant not smelling of fenugreek:
  3. Leaves alternate, or at least without enlarged nodes:
   4. Capsule smooth, sometimes densely covered with brightly coloured (red or orange) granules (§ *Rottlera*):
    5. Usually a climbing or scrambling shrub; ovary and capsule bilocular; flower-buds and fruit densely yellow-tomentose; leaves about as long as broad, membranaceous, slenderly nerved, with yellow granules beneath . . . . **M. repandus**
    5. Erect shrub or tree; ovary and fruit 3–4-locular;
     6. Fruit densely covered with carmine-red granules; leaves longer than broad, coriaceous, strongly trinerved at base with scalariform secondaries, closely red-granular beneath . . **M. philippensis**
     6. Fruit and leaves without red granules:
      7. Leaves less than twice as long as broad, often obtuse or rounded at apex . . . . **M. nesophilus**
      7. Leaves at least twice as long as broad, always acute at apex **M. discolor**
   4. Capsule densely or sparsely echinate (§ *Mallotus*):
    8. Leaves rhombic-ovate, whitish beneath, with two conspicuous round macular glands at extreme base; capsule whitish, bearing relatively short, not crowded, conic-subulate processes. . **M. paniculatus**
    8. Leaves less rhombic, often dentate towards apex, not white beneath, with much less conspicuous basal glands; capsules larger, with longer, slender, pubescent, densely crowded processes forming a continuous woolly layer, the whole infructescence often cylindric **M. mollissimus**
  3. Leaves strictly opposite (but sometimes exceedingly unequal, or with one member of each pair stipuliform or even obsolete), from slightly enlarged nodes:
   9. Leaves penninerved, or the basal nerves not stronger or more conspicuous than the remainder:
    10. Leaves glandular-granular beneath; ♂ inflorescence not or scarcely congested; ♀ inflorescence racemose; pubescence sparse, not stellate (§ *Axenfeldia*) . . . . **M. resinosus**
    10. Leaves not or scarcely glandular-granular beneath; ♂ inflorescence short and congested; ♀ inflorescence candelabriform; pubescence denser, mostly stellate (§ *Rottleropsis*) . . . . **M. claoxyloides**
   9. Leaves palmately nerved, or the basal nerves stronger or more conspicuous than the remainder; nodes and internodes often distinctly flattened (§ *Rottleropsis*):
    11. Capsule smooth or with a few small warts, bilocular, golden-tomentellous; leaves deeply cordate at base, ovate-orbicular **M. didymochryseus**

11. Capsule echinate:
  12. Leaves orbicular-ovate or triangular-ovate, the ultimate nerves forming a characteristic dense grey-papillose areolation under the lens; plant always growing near the coast; inflorescence not congested . . . . **M. tiliifolius**
  12. Leaves broadly or narrowly elliptic, nerves not densely areolate as above; plant not typically of coastal situations; inflorescences very short and congested . . . . **M. claoxyloides**

Mallotus claoxyloides (*F. Muell.*) *Muell. Arg.* in Linnaea 34: 192 (1865) & in DC.: 972 (1866); Benth.: 140 (1873); Moore & Betche: 78 (1893); Bailey: 1447 (1902); Pax & Hoffm. vii: 155 (1914); Baker, Hardw. Austr.: 359 (1919); Anderson: 219 (1968); Airy Shaw in K.B. 20: 42 (1966) & 31: 392 (1976), *in clavi*, & in Muelleria 4: 232 (1980). Type: Q., 'Ad flumen Brisbane', *Hill & Mueller* (MEL).

Echinocroton claoxyloides F. Muell., Fragm. 1: 32 (1858).
Rottlera (*Plagianthera* ?) *affinis* Baill., Ét. Gén. Euphorb.: 424, Atlas: 36, t. 19, figs. 29–31 (1858). Syntypes: Sine loc., *Verreaux*, '*Leichart*' (P).
Echinus claoxyloides (F. Muell.) Baill. in Adansonia 6: 315 (1866).

var. **claoxyloides**

(*Echinus claoxyloides* var. *genuina* Baill., *l.c.* (1866)).
M. claoxyloides var. *glabratus* Domin: 888 [334] (1927), **synon. nov.** Type: Q., Port Mackay, *Dietrich* 524 (PR).

Q (SK, ? PC, WB, MO); **NSW** (NC)–New Guinea.

Shrub to 3 m high, in mixed notophyll rain-forest on soils derived from mixture of basalt and metamorphic rocks up to 660 m.

A dry, harsh, somewhat scruffy-looking shrub or small tree, scabrid with stellate hairs. Leaves opposite, ovate, obovate or elliptic, rather short-petioled. Inflorescences (male and female) greatly abbreviated, the males capitate-racemose, the females capitate-umbellate, but candelabriform-elongate in the fruiting stage. Capsule densely minutely stellate and rather closely and shortly echinate.

var. **cordatus** (*Baill.*) *Airy Shaw* in Muelleria 4: 232 (1980). Syntypes: Q., Moreton Bay, *F. Mueller* (P, MEL); Rockhampton, salt water creeks, *Dallachy* (P, MEL); N.S.W., Richmond River & Clarence River, *Beckler* (P, MEL).

Echinus claoxyloides var. *cordata* Baill., *l.c.* (1866).
Mallotus claoxyloides var. *macrophylla* Benth.: 141 (1873); Bailey: 1447 (1902). Type: Q., Rockingham Bay, *Dallachy* (K).

Q (CK, NK); **NSW**—Endemic.

Spreading shrub in rain-forest or in mixed open forest on granite hillside up to 250 m.

A luxuriant, large-leaved form, softly tomentellous in all parts except the upper leaf-surface.

var. **ficifolius** (*Baill.*) *Benth.*: 141 (1873); Bailey: 1447 (1873); Britten, Ill. Austr. Pl. Cook's Voy.: 89, t. 292 (1905); Airy Shaw in Muelleria 4: 232 (1980). Type: Q., Rockhampton, *Dallachy* 47 (P, MEL).

Echinus claoxyloides var. *ficifolia* Baill. in Adansonia 6: 315 (1866).
Mallotus ficifolius (Baill.) Pax & Hoffm. vii: 151 (1914).
M. claoxyloides f. *grossedentata* Domin: 888 [334], *in obs.* (1927), **synon. nov.** Type: Q., Port Mackay, *Dietrich* 1834 (PR).

Q (CK, SK, PC)—Endemic.

Shrub or small tree, in dry bouldery creek-beds or amongst stones, or in gallery forest, at 100 m.

Doubtfully distinct from var. *cordatus*, but leaves broadly orbicular-ovate, up to 13·5 cm broad, and conspicuously repand-dentate; peduncle of infructescence up to 6 cm long.

var. **angustifolius** *F. M. Bailey*, Contrib. Queensl. Flora, in Queensl. Dept. Agric., Bull. No. 7 [Botany Bulletin No. 2]: 18 (1891), & Queensl. Fl.: 1447 (1902); Airy Shaw in Muelleria 4: 232 (1980). Type: Q., Yandina, *Bailey* (BRI).

Q (MO)—Endemic.

No ecological information.

A rather distinct form, with very shortly petioled (1–3 mm), narrowly cuneate-elliptic or oblanceolate, distantly pungent-repand-dentate leaves, and long-peduncled (to 4 cm) male inflorescences; capsules shortly and sparsely echinate. The mature leaves are glabrous except for minute scattered stellate hairs beneath. A similar form, but with almost entire leaves, was collected by Alan Cunningham in 1820 on Lizard Island (*Cunningham* 106).

MALLOTUS DERBYENSIS *W. V. Fitzg.* in Journ. Roy. Soc. W. Austr. 3: 165 (1918); Gardner: 72 (1931) = **Grewia** cf. **breviflora** *Benth.*, Fl. Austr. 1: 270 (1863); cf. Airy Shaw in Muelleria 4: 235 (1980).

Mallotus didymochryseus *Airy Shaw* in K.B. 20: 40 (1966) & 25: 527 (1971) & 31: 392 (1976). Type: Papua, Mt Lawes Forest, 1960, *K. J. White* NGF 8255 (K).

[*Croton arnhemicus* sensu Specht: 250, 382, 398, 461 (1958), *non* Muell. Arg.]

NT (DG)—New Guinea.

Shrub 2 m tall and wide, in area approaching monsoon forest on coastal dune.

From the other Australian members of the section *Rottleropsis* (leaves opposite and palminerved), *M. didymochryseus* differs in its racemose, up to 14-flowered ♀ inflorescence and especially in its much larger, dicoccous, smooth, densely orange-brown-stellate-tomentose capsule. The leaves are mostly very broadly orbicular-ovate, up to 25 cm diam., deeply cordate or occasionally rounded at the base. The male inflorescence (not previously described) is simply racemose, up to 20 cm long, fulvous-stellate-tomentose, the flowers in clusters, very shortly pedicelled; sepals 4, reflexed at anthesis;

stamens ± 100, 3–4 mm long. (Described from *Eddowes & Kumul* NGF 13094, Brown River, Papua.) Infructescence up to 10 cm long, with up to 14 didymous, smooth, fulvous-stellate-tomentose capsules, 17 × 13 × 13 mm.

In Australia this distinct species is so far known only from Groote Eylandt, in the Gulf of Carpentaria, where it was once collected by the Australian-American Expedition to Arnhem Land (*Specht* 437). The leaves of this specimen are less than 10 cm in diameter.

Mallotus discolor *F. Muell. ex Benth.*: 143 (1873); Moore & Betche: 78 (1893); Bailey: 1449 (1902); Pax & Hoffm. vii: 183 (1914); Baker, Hardw. Austr.: 362 (1919); Anderson: 219 (1968); Francis: 230 (1970); Airy Shaw in Muelleria 4: 232 (1980). Type: N.S.W., Clarence River, 1862, *London Exhibition* 82 (K).

Rottlera discolor F. Muell. in Coll. Northern Woods N.S.W., London Exhib. no. 82 (1862), *nomen.*
[*Macaranga mallotoides* sec. F. Muell., Fragm. 4: 140 (1864), *in obs.*, *non* F. Muell. 1863.]
[*Mallotus repandus* sec. F. Muell., Fragm. 6: 185 (1868), *in obs.*, *non* (Willd.) Muell. Arg.]

Q (NK, SK, PC, MO); **NSW** (NC)—Endemic.

Tree to 9 m tall, on frontal dunes or in rain-forest up to 30 m.

Very close to *M. nesophilus*, but the leaves at least twice as long as broad, and always acute at the apex. *M. discolor* occupies an entirely different geographical area from *M. nesophilus*, extending from northeast New South Wales north to the Elliot River, west of Bowen, 19° 5′ S, with a doubtful record from the Palm Islands, off Ingham, 18° 30′ S.

Mallotus mollissimus (*Geisel.*) *Airy Shaw* in K.B. 26: 297 (1971) & 31: 391 (1976) & K.B. Add. Ser. IV: 165 (1975) & Muelleria 4: 233 (1980). Type: 'Habitat in China' (no collector's name) (HAL?).

Croton mollissimus Geisel., Croton. Monogr.: 73 (March 1807).
C. ricinoides Pers., Syn. 2: 586 (Sept. 1807). Type: 'Hab. in India' (no collector's name) (P).
Mallotus ricinoides (Pers.) Muell. Arg. in Linnaea 34: 187 (1865) & in DC.: 963 (1866); Benth.: 139 (1873); Bailey: 1446 (1902); Pax & Hoffm. vii: 170 (1914); Francis: 230 (1970); Hyland: 60 (1971).
Echinus mollissimus (Geisel.) Baill. in Adansonia 6: 316 (1866).

Q (CK, NK, PC)—SE Asia to New Guinea and Melanesia.

Tree to 10 m high, in rain-forest up to 800 m.

Differs from *M. paniculatus*, the only other Australian member of *Mallotus* (leaves alternate, palminerved, capsule echinate), in the thinly floccose-tomentose indumentum of the triangular-ovate leaves, which are often cordate at the base, and in the densely arranged, slender but woolly, aculei of the capsule. The indumentum of the young parts is strongly cinnamomeous.

The species is scattered thinly from the Cape York Peninsula as far south as the region of Shoalwater Bay, north of Rockhampton.

Mallotus nesophilus *Muell. Arg.* in Linnaea 34: 196 (1865) & in DC.: 981 (1866); Benth.: 143 (1866); Bailey: 1449 (1902); Pax & Hoffm. vii: 183 (1914); Ewart & Davies: 167 (1917); Gardner: 72 (1931); Specht: 252 (1958); Airy Shaw in Muelleria 4: 233 (1980). Syntypes: Q., Cape Flinders, 1819, *Cunningham* 295 (G–DC); Sweers I., *Henne* (G–DC); Quail I., 1855, *Flood* (G–DC).

Echinus nesophilus (Muell. Arg.) Baill. in Adansonia 6: 314 (1866).

WA (K); **NT** (DG); **Q** (CK)—Endemic.

Shrub or tree to 8 m high, in mixed shrub woodland with scanty ground cover, at edge of monsoon forest on stable coastal dunes, in mixed open forest on sandy outwash plain, in thin rain-forest on sandy levee soil, or in sand in beach shrubland above high water mark, up to 125 m.

Closely related to *M. discolor*, but leaves less than twice as long as broad, often obtuse or rounded at the apex. *M. nesophilus* extends across the north of the continent from the Kimberleys to the Cape York Peninsula, where, however, it does not reach further south than the Chester River, in latitude 13° 40′ S.

Mallotus oblongifolius (*Miq.*) *Muell. Arg.* in Linnaea 34: 187 (1865) & in DC.: 962 (1866); Pax & Hoffm. vii: 193 (1914); Airy Shaw in K.B. 26: 306 (1971) & 31: 392 (1976), *in obs.*, & K.B. Add. Ser. IV: 173 (1975) & Muelleria 4: 233 (1980), *q.v.* Type: Java, near Tjikoja, *Zollinger* 245 (P).

Rottlera oblongifolia Miq., Fl. Ind. Bat. 1(2): 396 (1859).

Q (CK)—Andamans, SE Asia and throughout Malesia to New Guinea.

Shrub or tree by river at low altitude.

The claim of *Mallotus oblongifolius* to a place in the Australian flora rests upon a single oldish specimen, without flower or fruit, which I found in the Melbourne herbarium in 1976. It had been gathered on the Johnstone River in 1882 by a collector named *Berthoud*, about whom I have no further information. Collectors in the region of Innisfail should keep a look-out for this rather inconspicuous species, which might possibly persist in the area.

M. oblongifolius is one of the commonest and most widespread species of *Mallotus* in South-East Asia and Indonesia. It is the first member of the small section *Stylanthus* (leaves 3(–5)-nerved at base, opposite and/or alternate on the same branch; female calyx spathaceous; plant often smelling of fenugreek when dried) to be recorded from Australia. The leaves of *M. oblongifolius* are oblong to ovate, up to 9 × 19 cm, mostly shallowly cordate at base, shortly acuminate, shallowly denticulate or subentire, glabrous or shortly puberulous beneath and sparsely glandular-granular; petioles slender, very variable in length. Inflorescences slender, 1–15 cm long, minutely stellate-puberulous. Ovary densely muricate; styles 5–8 mm long, connate for $\frac{1}{2}$–$\frac{2}{3}$. Capsule up to 11 mm diam., echinate, densely granular-glandular.

Mallotus paniculatus (*Lam.*) *Muell. Arg.* in Linnaea 34: 189 (1865) & in DC.: 965 (1866); Benth.: 140 (1873); Bailey: 1446 (1902); Backer & Bakh. f.: 483 (1963); Airy Shaw in K.B. 26: 298 (1972) & K.B. Add. Ser. IV: 166 (1975). Type: Java, *Sonnerat* (P).

Croton paniculatus Lam., Encycl. Méth., Bot. 2: 207 (1786).
Mallotus cochinchinensis Lour., Fl. Cochinch.: 635 (1790); J. J. Sm.: 413 (1910); Pax & Hoffm. vii: 166 (1914). Type: 'Habitat sepes, et hortos minus cultos Cochinchinae, & Chinae', *Loureiro* (BM).

Q (CK, NK, SK)—SE Asia, S China, Formosa and throughout Malesia to New Guinea.

Tree of 6–9 m, in rain-forest, mostly on red basaltic soil, at 300–600 m.
Differs from *M. mollissimus*, the other Australian member of sect. *Mallotus*, in the rhombic-ovate, acuminate, mostly cuneate-based leaves, with a pair of conspicuous macular glands at the extreme base; in the exceedingly short, close, whitish tomentellum of the lower leaf-surface, and in the shorter, stouter, more conical, more distant aculei of the capsules. The indumentum of the upper surface of the young leaves is a deep madder brown.
I have not yet seen material of this species from farther north than the Mowbray River, south of Port Douglas.

Mallotus philippensis (*Lam.*) *Muell. Arg.* (corr. Merr. 1923) in Linnaea 34: 196 (1865) & in DC.: 980 (1866); Benth.: 141 (1873); Moore & Betche: 78 (1893); Bailey: 1447 (1902); J. J. Sm.: 450 (1910); Pax & Hoffm. vii: 184 (1914); Baker, Hardw. Austr.: 360 (1919); Merr.: 435 (1923); Backer & Bakh. f.: 484 (1963); Anderson: 218, 378 (1968); Airy Shaw in K.B. 21: 392 (1968) & 26: 300 (1972) & K.B. Add. Ser. IV: 168 (1975); Francis: 230 (1970); Hyland: 26 (1971). Type: Philippines, *Sonnerat* (P).

Croton philippense (!) Lam., Encycl. Méth., Bot. 2: 206 (1786).
Echinus philippinensis [Lam.] Baill. (*orth. mut.*) in Adansonia 6: 314 (1866). Type as above.

NT (DG); **Q** (CK, NK, SK, PC, WB, MO); **NSW** (NC)—W Himalaya and Ceylon to Formosa, and throughout Malesia to Melanesia.

Tree to 14 m high, in rain-forest or monsoon forest, in hilly Eucalypt country, on volcanic alluvial soil or brown silty loam or gravelly shale, up to 800 m.
The crimson glandular granules on the lower leaf-surface, and especially on the capsules, immediately distinguish *M. philippensis* from the other members of sect. *Rottlera* (*M. repandus*, *M. nesophilus*, *M. discolor*; leaves alternate, palmi- or trinerved; capsule smooth). The strongly trinerved leaves often somewhat recall those of *Cinnamomum*; the secondary venation is conspicuously transverse-scalariform. The indumentum of the young parts is mostly ferrugineous.

Mallotus polyadenos *F. Muell.*, Fragm. 6: 184 (1868); Benth.: 142 (1873) ('*polyadenus*'); Bailey: 1448 (1902); Pax & Hoffm. vii: 198 (1914); Airy Shaw in K.B. 20: 43 (1966); Hyland: 23 (1971). Type: Q., Rockingham's Bay, *Dallachy* (MEL).

Q (CK, NK)—New Guinea.

Shrub or tree to 25 m tall, in rain-forest and fringing forest or dry rain-forest, or mixed mesophyll vine forest, or sclerophyll woodland, on red basaltic soils or metamorphic rocks and poor soils, or on a red stony hill, up to 760 m. Fig. 5B.

Mallotus polyadenos, the type (and only Australian) species of sect. *Polyadenii*, is readily identifiable from the presence of glandular granules on the upper leaf-surface as well as the lower. The cuneate-obovate or cuneate-elliptic leaves are mostly opposite, but often also alternate on the same shoot, as in sect. *Stylanthus*; they are sometimes shortly cordate at the base. The nervation is principally pinnate, but there is much variation, the basal pair of nerves being sometimes quite strong and extending half-way up the lamina. The capsules are very deeply tricoccous, with smooth subglobose cocci; dehiscence appears to be septicidal only, so that each seed remains enclosed in the caducous loculus.

A rather frequent species from Cape York south to the latitude of Ingham.

Mallotus repandus (*Willd.*) *Muell. Arg.* in Linnaea 34: 197 (1865) & in DC.: 981 (1866); Benth.: 142 (1866); Bailey: 1449 (1902); J. J. Sm.: 455 (1910); Pax & Hoffm. vii: 181 (1914); Backer & Bakh. f.: 483 (1963); Airy Shaw in K.B. 26: 301 (1971) & Muelleria 4: 234 (1980). Type: S India, *Klein* (B†).

Croton repandus Willd. in Neue Schrift. Naturf. Freunde Berlin 4: 206 (1803).

Q (CK, NK)—W Peninsular India & Ceylon to Formosa & Philippines, and scattered through Malesia to New Guinea and New Caledonia.

Straggling vine or shrubby tree climbing over other trees, often in beach forest, or in rain-forest up to 760 m.

The alternate, palmately nerved, sometimes narrowly peltate leaves and smooth capsule place this species in sect. *Rottlera.* It differs from the other Australian members of the section, *M. discolor* and *M. nesophilus*, in the climbing or scrambling habit (almost unique in the whole genus), and in the bilocular ovary and capsule. The indumentum is usually strongly yellow-ochraceous.

The species has been noted in scattered localities from the Three Islands group, between Capes Flattery and Bedford, from the Atherton Tableland, and south to Bowen and Proserpine.

Mallotus resinosus (*Blanco*) *Merr.*, Sp. Blanco.: 222 (1918) & Enum. 2: 436 (1923); Airy Shaw in K.B. 26: 294 (1971) & 31: 392 (1976) & K.B. Add. Ser. IV: 161 (1975) & Muelleria 4:234 (1980). Type: Philippines, Luzon, Pasig & Angono, near Manila, *Blanco.*

Adelia resinosa Blanco, Fl. Filip. ed. 2: 562 (1845); Muell. Arg.: 731 (1866).
Claoxylon muricatum Wight, Ic. Pl. Ind. Or. 5: 24, t. 1886 (1852) (var.?). Syntypes: S India, Travancore, Courtallam, *Wight* 2614 & 2672 (CAL?).
Mallotus muricatus (Wight) Muell. Arg. in Linnaea 34: 191 (1865) & in DC.: 972 (1866), *p.p.*; J. J. Sm.: 428 (1910); Backer & Bakh. f.: 484 (1963).
M. walkerae Hook. f., Fl. Brit. Ind. 5: 437 (1887). Type: Ceylon, *Mrs Walker* (K).
M. muricatus var. *walkerae* (Hook. f.) Pax & Hoffm. vii: 190 (1914).

Q (CK)—Indomalaya to New Guinea, but absent from Sumatra, Malaya (except N) and Borneo (except N).

Shrub or small tree in dry rain-forest up to 300 m.

Mallotus resinosus is the sole Australian representative of sect. *Axenfeldia* (leaves cuneate-obovate, penninerved, strictly opposite, glabrous, closely granular-glandular beneath but not above; internodes flattened in alternate planes; capsule echinate). It has so far been collected only at Cooktown, at Altonmoui at the southern end of the Cape York Peninsula, and on Banks (Moa) Island in the Torres Strait. In Southeast Asia and Malesia it appears to be strongly calcicolous.

Mallotus tiliifolius (*Bl.*) *Muell. Arg.* in Linnaea 34: 190 (1865) & in DC.: 969 (1866); J. J. Sm.: 443 (1910); Pax & Hoffm. vii: 148 (1914); Backer & Bakh. f.: 484 (1963); Airy Shaw in K.B. 26: 305 (1971) & 31: 392 (1976), *in obs.*, & K.B. Add. Ser. IV: 170 (1975) & Muelleria 4: 234 (1980). Type: Java, 'ad littora insularum Nusae Kambangae et Javae', *Blume* (BO).

Rottlera tiliifolia Bl., Bijdr.: 607 (1825).
Croton enantiophyllus K. Schum. in K. Schum. & Lauterb., Nachtr. Fl. Deutsch. Schutzgeb. Südsee: 296 (1905). Type: 'Bismarck Archipel: Neu-Pommern (R. Parkinson)' (B†).

Q (CK)—Formosa, Lower Siam, and throughout Malesia to Fiji.

Shrub or tree to 7·5 m tall, in mangroves and adjacent paper-bark swamps, or in sandy soil near beach, at very low altitudes.

In sect. *Rottleropsis*, this species is distinguished from all others by the densely reticulate grey-papillose undersurface of the leaves, which is closely foveolate with the small pits in which glandular granules are sunk. The minute stellate indumentum of the whole plant has a dry, dusty, powdery appearance, with a pale cinnamomeous tinge on the young parts. *M. tiliifolius* never grows far from the sea. In Australia it is so far known only from Trinity Beach, north of Cairns, and from Prince of Wales Island in the Torres Strait.

Margaritaria *L. f.*

Closely related to *Securinega*, but differing in its mostly arborescent habit (up to 30 m tall) and deciduous leaves; sepals 4, stamens 4; no pistillode; ovary 2–3-locular; fruits mostly larger, non-fleshy, on longer pedicels, ripe seeds with a metallic blue testa.

1. Leaves thinly chartaceous or submembranaceous, acute, blackish-green when dry, margin finely undulate; ovary and fruit trilocular, 8–11 mm diam., pedicel 10–15 mm; in rain-forest . . . . **M. indica**
1. Leaves coriaceous, obtuse to emarginate, yellow-veined when dry, margin not undulate; ovary and fruit bilocular, 5–7 mm diam., pedicel 4–8 mm; in dry open forest . . . . **M. dubium-traceyi**

Margaritaria dubium-traceyi *Airy Shaw & Hyland* in K.B. 31: 357 (1976) & in Muelleria 4: 214 (1980). Type: Q., Dimbulah–Petford, 1974, *Hyland* 7937 (K).

Q (BK, CK)—Endemic.

Shrub or small tree to 5 m tall, with smooth white-and-grey piebald bark, in dry open forest in sandstone areas or among granite boulders, up to 600 m.

In its coriaceous leaves, obtuse or rounded or minutely emarginate at the apex and distinctly yellow-veined when dry, in its small bilocular fruits, 5–7 mm diam., borne on short (4–8 mm) pedicels, and in its dry open forest habitat, this is very distinct from *M. indica.* The epithet is a rough rendering of 'Tracey's Puzzle', the name by which the plant came to be known when Mr Geoff Tracey, of C.S.I.R.O. Brisbane, sent material for naming to the Forest Research Station, Atherton. The staff were unable at that time even to suggest a genus for it, and the name indicates the perplexity which it caused them.

Though only described within the last few years, this species was first collected just 100 years ago, by Armit on the Robertson River. It is now known from an area extending from a point near the Queensland–Northern Territory border, southwest of Burketown, eastwards to Forsayth and the Chillagoe–Petford region, and thence northwards as far as Laura.

Margaritaria indica (*Dalz.*) *Airy Shaw* in K.B. 20: 387 (1966) & 26: 308 (1971) & 31: 356 (1976) & K.B. Add. Ser. IV: 175 (1975). Type: No material explicitly cited; India, Concan, *Dalzell* (K), ex herb.

Prosorus indicus Dalz. in Hook. Journ. Bot. & Kew Garden Misc. 4: 346 (1852).

Q (CK)—India and Ceylon to Formosa, SE Asia, and Malaya (v. rare), and scattered through Malesia to New Guinea.

Tree to 15 m high, in gallery rain-forest at 30–80 m.

The thinner, chartaceous or submembranaceous leaves, with a very acute apex, drying blackish-green, with a very finely undulate margin and reddish or concolorous nerves, the considerably larger trilocular fruit (8–11 mm diam.) on 10–15 mm long pedicels, and the gallery rain-forest habitat, distinguish *M. indica* sharply from *M. dubium-traceyi.* The species, though so widely distributed in Indomalesia, appears to be restricted in Australia to a small area near Iron Range in the Cape York Peninsula.

Neoroepera *Muell. Arg. & F. Muell.*

Shrubs or small trees; leaves small, alternate, entire, chartaceous, very shortly petioled; stipules minute or obsolete. Flowers in axillary fascicles, the females solitary, either alone or in a male fascicle. Male flower: sepals (2–)3 + 3, free, broadly elliptic, ± obtuse; stamens 5–6, free, surrounding a broad central disk. Female flower: sepals 3 + 3, narrowly elliptic, acute, very unequal; disk shortly lobed; ovary 3-locular, loculi biovulate; stigmas subsessile or stipitate, flabellate or clavate-capitate. Capsule tricoccous, seeds subcylindric or ellipsoid, smooth and shining; embryo with broad (1·7 mm) cotyledons (*teste* Bentham). An isolated genus, perhaps distantly related to *Kairothamnus*, of New Guinea.

1. Monoecious; leaves elliptic, up to 3·8 × 1·8 cm; fascicles many-flowered; ♂ pedicels 10–12 mm long; stamens exserted . . . . . **N. buxifolia**
1. Dioecious; leaves cuneate-oblanceoate, up to 2·3 × 0·7 cm; fascicles few-flowered; ♂ pedicels 5–7 mm long; stamens scarcely exserted **N. banksii**

Neoroepera banksii *Benth.*: 117 (1873); Bailey: 1425 (1902); Airy Shaw in Muelleria 4: 217 (1980). Type: Q., Endeavour River, north shore, *Cunningham* (K).

Q (CK)—Endemic.

Shrub to 2 m high, abundant in messmate and bloodwood savanna-forest of lateritic ridges, on sandy ridges by river, in heathy open forest, in low open shrubland on low dunes, or in closed scrub on high dunes, up to 40 m. Fig. 5A.

Stems papillose-puberulous. Leaves cuneate-oblanceolate or scarcely cuneate-obovate, sometimes almost oblong, 1–2·3 × 0·3–0·7 cm, very shortly petioled, rounded or emarginate or mucronulate at the apex, stiffly chartaceous, glabrous. Flowers dioecious, the males in few-flowered fascicles, the females solitary. Male flowers: pedicel 5–7 mm; sepals scarious, shining, the outer scarcely 1 mm, the inner 1·5 mm long; stamens equalling the inner sepals, scarcely exserted. Female flowers: pedicel stouter, 1·5–2·5 cm long; sepals herbaceous, the outer scarcely 1 mm, the inner 2 mm long; ovary glabrous, with 3 subsessile or stipitate (1 mm), suborbicular or flabellate stigmas. Capsule 6–7 mm diam.; seeds shortly subcylindrical, 5 × 2·5–3 mm, very smooth and shining, chestnut or deep brown.

The species seems to occur in scattered localities from Cape York to Cooktown.

Neoroepera buxifolia *Muell. Arg. & F. Muell.* in DC.: 489 (1866); Benth.: 116 (1873); Bailey: 1425 (1902); Pax & Hoffm. in Engl. & Harms, Nat. Pflanzenf. ed. 2, 19c: 73 (1931); Airy Shaw in Muelleria 4: 218 (1980), *in obs.* Type: Q., Princhester Creek, *Bowman* (MEL).

Securinega muelleriana Baill. in Adansonia 6: 333 (1866), *pro nom. nov.*; non *S. buxifolia* (Poir.) Muell. Arg. Type as above.

Q (PC)—Endemic.

Small tree, locally common along creeks at low altitude.

Differs from *N. banksii* in its monoecious habit, elliptic leaves, up to 3·8 × 1·8 cm, many-flowered fascicles (either all male or with a single female also), longer male pedicels (10–12 mm), slightly longer inner sepals, distinctly exserted stamens, narrower, clavate stigmas, and broader ellipsoid seeds, almost black when ripe.

Apparently restricted to a small area between Marlborough and Princhester, north of Rockhampton.

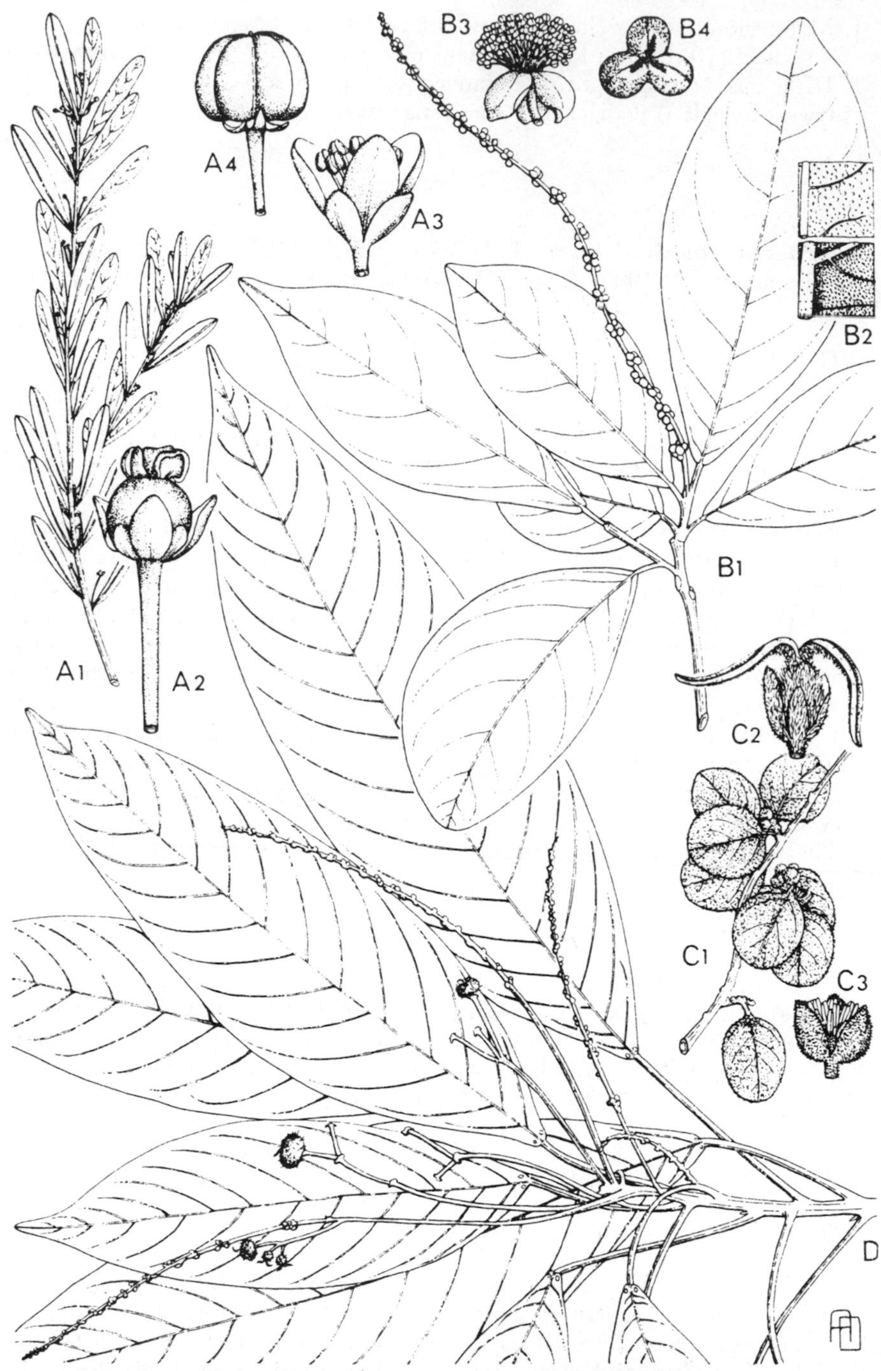

FIG. 5. *Neoroepera banksii:* **A1** habit × ⅔, from *Brass* 18823; **A2** female flower × 3⅓, from *Brass* 18822; **A3** male flower × 8, from *Brass* 18823; **A4** fruit × 2, from *McDonald & Batianoff* 1619. *Mallotus polyadenos:* **B1** habit × ⅔, from *Smith* 12404; **B2** leaf surfaces × 2, from *Smith* 12404; **B3** male flower × 2⅔, from *Hyland* 3679; **B4** fruit × 1⅓, from *Hyland* 7719. *Petalostigma nummularium:* **C1** habit × ⅔, from *Lazarides* 5866; **C2** female flower × 2, from *Allen* 677; **C3** male flower × 2, from *Lazarides* 5866. *Macaranga subdentata:* **D** habit × ⅔, from *Smith* 11121.

Omphalea *L.*

Mostly heavy lianas, or erect shrubs when growing in open places. Leaves alternate, entire or lobed, penninerved or palmatinerved, coriaceous, glabrous; petioles moderate, conspicuously biglandular at the apex; stipules small and triangular or obsolete. Inflorescences terminal, thyrsoid, bisexual, the flowers arranged in small cymes consisting of several males surrounding a single central female, each cyme subtended by a large, elongate-oblong or elliptic, membranaceous, pale-coloured, sometimes long-petioled bract, the petiole often biglandular at the apex. Male flower: sepals 4–5, orbicular-ovate, much imbricate; petals 0; disk-glands 5, fleshy, inserted at the base of the sepals, or connate into a fleshy cone surrounding the staminal column; stamens 2–3, connate into a very short column, either shortly free at the apex or united into a peltate or pileiform connective, bearing 2–3 reniform anthers on the margin. Female flower (larger than the male): sepals as in male; disk 0; ovary 2–3-locular, loculi 1-ovulate; stylar column very thick, ovoid, entire, with a small terminal stigmatic pore. Fruit large or very large, subglobose, capsular and dehiscent or subdrupaceous and tardily dehiscent; seed large, trigono-subglobose, ? thinly arillate when fruit capsular.

Omphalea queenslandiae *F. M. Bailey* in Meston, Rep. Exped. Bellenden-Ker Range: 58 (1889); Bailey, Synopsis Queensl. Fl., Third Suppl.: 67 (1890) & Queensl. Fl. 5: 1455, t. 66 (1902); Pax & Hoffm. v: 19 (1912); Airy Shaw in K.B. 20: 415 (1966), *in obs.*, & 23: 130 (1969) & in K.B. Add. Ser. VIII: 177 (1980), *q.v.* for full synon. Type: Q., Johnstone River, *Bancroft* (BRI).

Q (CK)—New Guinea, Solomon Is.

Vigorous woody liane, climbing to 30 m in tops of canopy trees in rain-forest, up to 20 m.

Leaves triangular-ovate or elliptic-oblong, up to 30 cm long and 15 cm wide, cordate, truncate or cuneate at the 3–5-nerved base, rounded or cuspidate or shortly caudate at the apex; petiole 1–9 cm long, with 2 conspicuous sessile lateral glands at the apex. The plant starts life as a shrub, and the early leaves are often deeply 3- or even 5-lobed. Floral bracts narrowly linear, to 2·5 cm long. Fruits subglobose, to 6·5 cm diam., markedly 3-angled, orange; seeds roughly hemispherical, 2·5 × 2 × 1·5 cm, coarsely, but shallowly and irregularly, longitudinally ridged, black.

Petalostigma *F. Muell.*

Trees or shrubs, pubescent or subglabrous, hairs simple. Leaves alternate, simple, entire or shallowly crenate, shortly petioled. Flowers dioecious, in axillary fascicles, shortly pedicelled. Male flower: sepals (? bracts) 4–6, suborbicular, much imbricate; disk 0; stamens 20–40, borne on a shortly conical androphore often pilose below, anthers extrorse, often with a pilose connective; pistillode 0 (or exceptionally developed as an irregularly bilobed structure). Female flowers (mostly solitary or paired): sepals (? bracts) 4–8, the outer ± lanceolate, the inner ovate-orbicular; ovary 3–4-locular, loculi biovulate; styles large, variously flabellate or petaloid, undulate or lobulate, fleshy, caducous. Capsule globose, 3–4-locular, with a bright orange ± fleshy exocarp; seeds oblong-ellipsoid, glossy, mostly with a small conspicuous caruncle.

1. Anthers, ovary and capsule quite glabrous; leaves sometimes shallowly crenate; indumentum of lower leaf-surface sericeous, or at any rate hairs about 1 mm long; undershrub, less than 1 m high, often sending up vigorous unbranched shoots after burning . . . **P. quadriloculare**
1. Anthers (or at least connective), ovary and capsule ± pubescent; leaves entire (rarely minutely and obscurely crenulate); indumentum usually very finely and densely felted, hairs minute (± 0·5 mm long), or sometimes almost absent:
  2. Leaves spathulate or subspathulate (rarely suborbicular), often retuse, glabrous or almost so beneath; male flowers often precocious; staminal column glabrous below; tree to 4·5 m . . . . . **P. banksii**
  2. Leaves elliptic or ovate or orbicular, always ± pubescent or pilosulous beneath; male flowers rarely precocious; staminal column usually pilose below:
    3. Leaves very stiffly coriaceous, the upper side with a finely shagreened or minutely granular surface; shrub of 1–2 m . **P. pachyphyllum**
    3. Leaves chartaceous or subcoriaceous, the upper surface not shagreened or granular:
      4. Leaves rather narrowly elliptic, to 8 cm long, acute at apex and often at base also, usually drying dark brown above; capsule puberulous or glabrous, mostly trilocular; small tree . . . . **P. triloculare**
      4. Leaves shorter and broader in proportion, ovate, orbicular or shortly elliptic, often obtuse at apex, rarely drying dark brown above; capsule pubescent, mostly 4-locular:
        5. Leaves orbicular or suborbicular, 1–2·5 cm long, densely grey-tomentose, especially below; shrub of 1–2 m . **P. nummularium**
        5. Leaves ovate or shortly elliptic, 2–3(–5) cm long, shortly pubescent; tree to 7·5 m . . . . **P. pubescens**

Petalostigma banksii *Britten & S. Moore* in Journ. Bot. Brit. & For. 41: 225, t. 453 (1903); Britten, Illustr. Austr. Pl. Banks & Soland. 3: 87, t. 286 (1905); C. T. White in Queensl. Nat. 1: 208 (1911); Bailey in Queensl. Agric. Journ. 27: 68 (1911) & Compreh. Cat. Queensl. Pl.: 477, fig. 465 (1913); Pax & Hoffm. xv: 283 (1922); Domin: 871 [317] (1927); Blake in Austr. Journ. Bot. 2: 108 (1954); Chippendale in Proc. Linn. Soc. N.S.W. 96: 245 (1972): Airy Shaw in K.B. 31: 369 (1976). Type: Q., Endeavour River, 1790, *Banks & Solander* (holotype BM).

NT (DG); **Q** (BK, CK, NK, SK)—Endemic.

Small tree to 4·5 m tall, in savanna-forest of sandy granitic ridges, on granite outcrops, on rocky conglomerate slopes in *Acacia* forest, on red soil or sandy yellow or red lateritic podsol or gravelly red earth in *Eucalyptus* woodland, up to 450 m.

Distinguished by its arborescent habit, spathulate (rarely suborbicular) leaves, often retuse at the apex and almost glabrous beneath, and its sometimes precocious male flowers, with a glabrous staminal column. The plant seems to have a rather sparse and scattered distribution.

Petalostigma nummularium *Airy Shaw* in K.B. 31: 373 (1976). Type: N.T., W of Frewena, 1971, *Dunlop* NT 31094 (K).

P. quadriloculare var. *nigrum* Ewart & Davies: 166, t. 17 (1917); Chippendale in Trans. Roy. Soc. S. Austr. 83: 200 (1960), *q.v.*, & in Proc. Linn. Soc. N.S.W. 96: 245 (1972). Syntypes: N.T., Borroloola, *G. F. Hill* 908; N of 15 degrees, *Campbell* 17; 70 miles N of Camp IV, *G. F. Hill* 387 (all MEL).

WA (K); **NT** (DG, BT)—Endemic.

Shrub of 1–2 m, in dense or open scrub on deep red sand or in coarse clayey sand, or on red loam and lateritic deposits; altitudes not stated. Fig. 5C.

The shrubby habit and the orbicular densely grey-tomentose leaves, not exceeding 2·5 cm in length, are the distinguishing features of this plant, but some dwarf forms of *P. pubescens* approach it rather closely.

The distribution is centred on an area surrounding Tennant Creek, in the centre of the Northern Territory, with an outlier at a similar latitude (19°–20° S) just across the border in Western Australia.

Petalostigma pachyphyllum *Airy Shaw* in Kew Bull. 31: 372 (1976). Type: Q., Rockland Springs, 1961, *Lazarides & Story* 120 (holotype K).

Q (LT)—Endemic.

Shrub of 1–2 m, abundant and locally dominant in thin sandy soil on dissected plateau of Triassic sandstone, with shrub woodland, or on a sandy and stony ridge, or on dry grey shallow sandy soil over sandstone, at 600–900 m.

The shrubby habit, and the stiffly coriaceous leaves with a finely shagreened upper surface, often turning a pinkish-brown colour when dry, with relatively inconspicuous nervation, distinguish *P. pachyphyllum* from the remaining species. It seems to be a very local plant, restricted to a limited area in Leichhardt district, between Emerald and Injune.

Petalostigma pubescens *Domin*: 871 [317] (1927); Airy Shaw in K.B. 29: 303 (1974) & 31: 368 (1976); George & Kenneally in Miles & Burbidge, Biol. Surv. Prince Regent River Reserve: 47 (1975) & in Kabay & Burbidge, Biol. Surv. Drysd. River Nat. Park: 54 (1977); Airy Shaw in K.B. Add. Ser. VIII: 178 (1980). Lectotype: Q., 'Sub-Tropical New Holland', 1846, *Mitchell* 615 or 438, young fruiting specimen (K) (see K.B. 31: 368 (1976)).

P. quadriloculare β pubescens Muell. Arg. in Flora 47: 481 (1864); Pax & Hoffm. xv: 282 (1922); Chippendale in Proc. Linn. Soc. N.S.W. 96: 245 (1972). Type: N.T., Arnhem's Land, 1855–7, *F. Mueller s.n.*, ♂ specimens (K).

P. quadriloculare β genuina Muell. Arg.: 273 (1866), *nom. illegit., superfl.* Type as for var. *β pubescens*.

P. australianum Baill. in Adansonia 7: 356, *pro majore parte*, t. 2, figs. 1–8 (1867), *nom. nov. illegit.* based on: 'Crescit in Australia boreali-australi [*sic*!], circa sinum Carpentariae, ubi detexit olim *Leichhardt* (herb. Mus. par.!)'.

[*P. quadriloculare* sec. F. Muell., Fragm. 6: 182 (1868), *saltem pro parte*; Moore & Betche: 73 (1893); Anderson: 217 (1968); *non* F. Muell. (1857).]

[*P. quadriloculare* var. *glabrescens* sec. Ewart & Davies: 166 (1917); Moore & Betche, *l.c.* (1893); Chippendale, *l.c.* (1972); *non* Benth.]

WA (K); **NT** (DG, VR, BT, CA); **Q** (BK, CK, NK, MI, LT, PC, BT, WB, MO); **NSW** (NC, NWS, NWP)—Papua.

Shrub or small tree to 6 m high, in scrubby open forest on rocky sandstone slopes, or on deep sandy soil or red sandy loam with lateritic outcrops, or on quartzite hill, or in coastal sand-dunes, or on granitic ridges, or on basalt, sometimes near rain-forest or mangroves, up to 900 m.

An exceedingly variable taxon, generally forming a small tree; recognizable more by a series of negatives than by positive features. It lacks the suffruticose habit, long silky indumentum, sometimes crenate leaves, and glabrous anthers, ovary and capsule, of *P. quadriloculare*; the more or less spathulate, often retuse leaves, glabrous beneath, and the glabrous staminal column, of *P. banksii*; the suffruticose habit and very stiffly coriaceous leaves, finely shagreened or minutely granular above, of *P. pachyphyllum*; the narrowly elliptic leaves, up to 8 cm long, acute at base and apex, of *P. triloculare*; and the suffruticose habit and orbicular or suborbicular, densely grey-tomentose leaves of *P. nummularium*.

The geographical distribution exceeds the combined distribution of all the remaining species.

Petalostigma quadriloculare *F. Muell.* in Hook. Journ. Bot. & Kew Garden Misc. 9: 17 (1857); Benth.: 92 (1873), *pro minore parte*; Ewart & Davies: 166 (1917), *pro parte*; Domin: 870 [316] (1927); Gardner: 72 (1931); Specht: 252, 382–3, 398, 461 (1958), *saltem pro parte*; Chippendale in Proc. Linn. Soc. N.S.W. 96: 245 (1972) (var. *quadriloculare*); Airy Shaw in K.B. 31: 366, *in obs.*, 378 (1976). Lectotype (1976, p. 366): N.T., Arnhem's Land, 1855–7, *F. Mueller* (MEL, ♀ material only).

Hylococcus sericeus R. Br. ex Mitchell, Journ. Exped. Int. Trop. Austr.: 389, 433 (1848), *nomen subnudum*; Britten & S. Moore in Journ. Bot. Brit. & For. 41: 226 (1903), *in obs.*

P. quadriloculare α *sericeum* Muell. Arg. in Flora 47: 481 (1864) & in DC.: 273 (1866); Chippendale, *l.c.* (1972). Type: 'In Arnhemsland Novae Hollandiae (*Dr Ferd. Müller*)' (G–DC).

P. australianum Baill. in Adansonia 7: 356 saltem quoad synon. *P. quadriloculare* F. Muell., p.p., (1867), *nom. nov. illegit.*

Xylococcus sericeus R. Br. ex Britt. & S. Moore, *l.c.* (1903), *in obs.*

P. humilis [sic] W. V. Fitzg. in Journ. Roy. Soc. W. Austr. 3: 163 (1918); Gardner: 72 (1931). Type: W.A., King River [nr. Wyndham], 1906, *Fitzgerald* (MEL).

[*P. quadriloculare* var. *pubescens* sec. Pax & Hoffm. xv: 282 (1922), quoad synon. var. *sericeum* Muell. Arg. tantum; *non* var. *pubescens* Muell. Arg.]

P. haplocladum Pax & Hoffm.: 283 (1922); Chippendale, *l.c.* (1972). Syntypes: N.T., north coast, *Schomburgk* (B†); Port Darwin, *Schultz* 447 & 449 (B†, K).

P. micrandrum Domin: 871 [317] (1927). Type: N.T., 'Port Heats' [Keats], Sept. 1819, *Cunningham* 477 (K).

Undershrub to 1 m high, common in open *Eucalyptus* forest on truncated yellow or lateritic podsol, or in damp areas on sandstone escarpment, or on grey sandy soil, or on undulating stony ground, or on slopes of a volcanic hill, or on upper slopes and crests of a granite outcrop; altitudes not stated.

Petalostigma quadriloculare is the most distinctive species in the genus, from its suffruticose pyrophytic habit, its silky indumentum, its sometimes shallowly crenate leaves, and its entirely glabrous stamens, ovary and capsule.

It is very characteristic of the Top End of the Northern Territory, extending westward only a short distance into Western Australia (east Kimberley region) and eastward just into the Burke district of Queensland.

Petalostigma triloculare *Muell. Arg.* in Flora 47: 471 (1864) & in DC.: 274 (1866); Airy Shaw in K.B. 31: 369 (1976). Type: Q., Moreton Bay, *F. Mueller* (holotype G-DC; isotypes K, MEL).

P. australianum Baill. in Adansonia 7: 356 (1867), *p.p.*, quoad synon. *P. triloculare* Muell. Arg., *nom. nov. illegit.*

P. quadriloculare var. *glabrescens* Benth.: 92 (1873); Bailey, Syn. Queensl. Fl.: 409 (1883) & Queensl. Fl. 5: 1428 (1902) & Compreh. Cat. Queensl. Pl.: 477, fig. 467 (1913); Pax & Hoffm. xv: 283 (1922). [*Non* Chippendale in Proc. Linn. Soc. N.S.W. 96: 245 (1972) = *P. pubescens* Domin.] Syntypes: Q., Moreton Bay, *W. Hill* (MEL, K); Cape Sidmouth, *Curdie* (MEL); N.S.W., Clarence River, *C. Moore* (MEL); London Exhibition, 1862, no. 91 (K).

Q (PC, WB, MO); **NSW** (NC)—Endemic.

Distinguished from the two other arborescent species (*P. pubescens* and *P. banksii*) by the narrowly elliptic leaves, up to 8 cm long, acute at the apex and often at the base also, and the mostly 3-locular ovary and fruit.

Restricted to a strip of country extending from NE New South Wales (29° 48′ S) about 300–350 km northwards at least as far as the region of Shoalwater Bay (22° 40′ S).

Phyllanthus *L.*

Trees, shrubs or herbs (sometimes annual), without stellate hairs. Leaves alternate, simple, entire, short-petioled, often distichously arranged along branchlets, simulating pinnate leaves; stipules various, sometimes peltate. Flowers small, mostly axillary, fascicled, sometimes in racemes or loose panicles, monoecious. Male flower: sepals 3 + 3, occasionally 2 + 2 or more than 6, free; disk-glands various, sometimes large and lacunose; stamens 2–6, filaments free or connate, anthers dehiscing vertically or transversely; pistillode 0 or rarely minute. Female flower: sepals much as in male; disk-glands mostly small, free, or connate into an annulus, or occasionally into an urceolate structure surrounding the ovary; ovary mostly 3-locular, loculi biovulate; styles free or shortly connate, bifid or rarely entire, spreading. Fruit capsular, rarely baccate or drupaceous, of mostly 3 bivalved cocci; seeds trigonous, testa smooth or sculptured or sometimes bearing hygroscopic trichomes, exarillate.

There are now perhaps 50 species of *Phyllanthus* known in Australia, the majority consisting of small undershrubs included by Bentham in his mixed-bag group *Paraphyllanthus*. Subgenus *Kirganelia* is represented by the same 3 species as in Bentham (*P. baccatus* F. Muell. ex Benth. = *P. ciccoides* Muell. Arg.). Subgen. *Isocladus* is represented by the variable complex centred around *P. virgatus* Forst. f., with a few other taxa of doubtful status that have been added later (by Spencer Moore, Domin, etc.). Subgen. *Eriococcus* (*Reidia*) is represented by *P. armstrongii* Benth. (possibly never re-collected since the time of Armstrong), and by the Malesian *P. lamprophyllus* Muell. Arg., discovered in recent years in the Cape York Peninsula. *P. brassii* C. T. White, endemic on Thornton Peak, evidently belongs in the same general affinity, but is a remarkably distinct plant. Subgen. *Gomphidium* § *Nymania* has now 3 Australian representatives, found also in New Guinea—*P. clamboides* (F. Muell.) Diels, *P. cuscutiflorus* S. Moore and *P. praelongipes* Airy Shaw & Webster. The recently described *P. sauropodoides* Airy Shaw (extraordinarily mimicking *Sauropus macranthus* Hassk.) has clear New Guinea relationships and is perhaps referable to *Gomphidium* § *Adenoglochidion*, a group founded on certain New Caledonian species. *P. hypospodius* F. Muell., with 5–6 free stamens and as many small flattened reniform glands at their base, may belong here also. The whole *Phyllanthus* group needs careful revision by a worker in Australia.

Pimelodendron *Hassk.*

Medium-sized or tall trees, glabrous. Leaves alternate, obovate or elliptic, distinctly or obscurely crenate, coriaceous, penninerved, moderately petiolate; stipules obsolete. Flowers dioecious. Male inflorescences shortly racemose, axillary or more often extra-axillary, mostly fascicled; flower very shortly pedicelled, in the axil of a small bract; calyx bivalved, consisting of 2 reniform sepals; stamens 10–12, the anthers subsessile. Female inflorescences shortly and loosely racemose, mostly solitary and axillary; flowers moderately (to 5 mm) pedicelled, bract minute; calyx shortly cupular, shortly 2–3-lobed; ovary shortly cylindric-oblong, crowned by the sessile pulvinate stigma slightly broader than the ovary. Fruit an ellipsoid drupe, contracted and bluntly 3–4-angled at base and apex, 2·5 cm long; seed solitary, broadly sessile, depressed-globose or subreniform, testa finely radiate-striolate from the apex, with a very finely reticulate surface, the seed apparently almost half enclosed in a very broad, shallow, multilobulate cupule (? aril).

Pimelodendron amboinicum *Hassk.* in Versl. & Meded. Akad. Wetensch. Amsterd. 4: 140 (1855) & ['*Pimeledendrum*'] Hort. Bogor. s. Retziae ed. nov.: 69 (1858); Pax & Hoffm. v: 54, fig. 9 (1912); Merr., Interpr. Rumph. Herb. Amboin.: 327 (1917); Whitmore, Guide For. Brit. Sol. Is.: 71 (1966); Airy Shaw in K.B. 31: 398 (1976) & in K.B. Add. Ser. VIII: 196 (1980). Type: Cult. in Hort. Bogor. (e planta a *W. Cleerens* ex insulis Moluccanis missa), 1854 (?), *Hasskarl* (BO).

Carumbium amboinicum (Hassk.) Miq., Fl. Ind. Bat. 1(2): 413 (1859); Muell. Arg.: 1143 (1866).

Pimeleodendron papuanum Warb. in Engl., Bot. Jahrb. 18: 198 (1893); **Pax** & Hoffm. v: 55 (1912). Syntypes: NE New Guinea, Finschhafen, *Hellwig* 403 & 464 (B†).

Daphniphyllum conglutinosum Hemsl. in Bull. Misc. Inf. Kew 1895: 137 (1895); Whitmore, *l.c.*: 175 (1966). Type: Solomon Is., San Cristoval, *Comins* 75 (K).

Q (CK)—Celebes, Moluccas, New Guinea (very abundant), Bismarcks, Solomon Is.

Tree to 25 m tall, in gallery rain-forest at 60–80 m. Fig. 6A.

Entirely glabrous. Leaves elliptic, 8–25 cm long, coriaceous, distantly and shallowly crenate (not unlike those of *Cleidion spiciflorum*), cuneate or sometimes rounded at base, cuspidate or shortly acuminate at apex; petiole slender, 1–8 cm long. Inflorescences shortly racemose, fascicled, axillary or extra-axillary, the males slender, several- to many-flowered, up to 4 cm long, the females stout, 1–6-flowered, up to 6 cm long. Male flowers very shortly pedicelled; calyx broadly cup-shaped, broadly 2-lobed; stamens 10–12, with subsessile anthers. Female flowers 2–3 mm pedicelled; calyx similar to male; ovary ellipsoid-obovoid, with a broad sessile pulvinate stigma. Drupe obovoid or subglobose, up to 5 × 3·5 cm, red when ripe, with a single seed in cream flesh.

The species is not yet known outside the Cape York Peninsula, where it appears to be very local.

RICINUS L.

RICINUS COMMUNIS *L.*, Sp. Pl.: 1007 (1753); Muell. Arg.: 1017 (1866); Hook. f.: 457 (1887); Bailey: 1452 (1902); J. J. Sm.: 537 (1910); Pax & Hoffm. ix–xi: 119 (1919); Merr.: 344 (1921) & 447 (1923); Corner: 274 (1940); Backer & Bakh.: 492 (1963); Airy Shaw in K.B. 26: 328 (1971), & K.B. Add. Ser. IV: 191 (1975), & K.B. Add. Ser. VIII: 198 (1980).

WA; **Q**; **NSW**; **SA**—Widely cultivated in all tropical countries; perhaps native in north-east tropical Africa (Somaliland, N Kenya, etc.), preferring stream-beds and soils with high nitrogen or saline content.

Shrub of 2–3 m, naturalized on roadsides and waste ground.

The castor oil plant, often grown as an ornamental.

Rockinghamia *Airy Shaw*

Distantly related to *Mallotus*, but differing in the pseudo-verticillate arrangement of the leaves, the divaricate-fasciculate-thyrsoid inflorescences, with male and female flowers intermingled, the presence of short, pilose, juxta-staminal processes amongst the stamens, and the closely muriculate capsules.

1. Petioles slender, up to 4·5 cm long; pistillode in male flower 0, or very small; styles bifid, segments slender, minutely papillose **R. angustifolia**
1. Petioles robust, up to 1·7 cm long; large pistillode in male flower; styles simple, thick, conspicuously laciniate . . . . **R. brevipes**

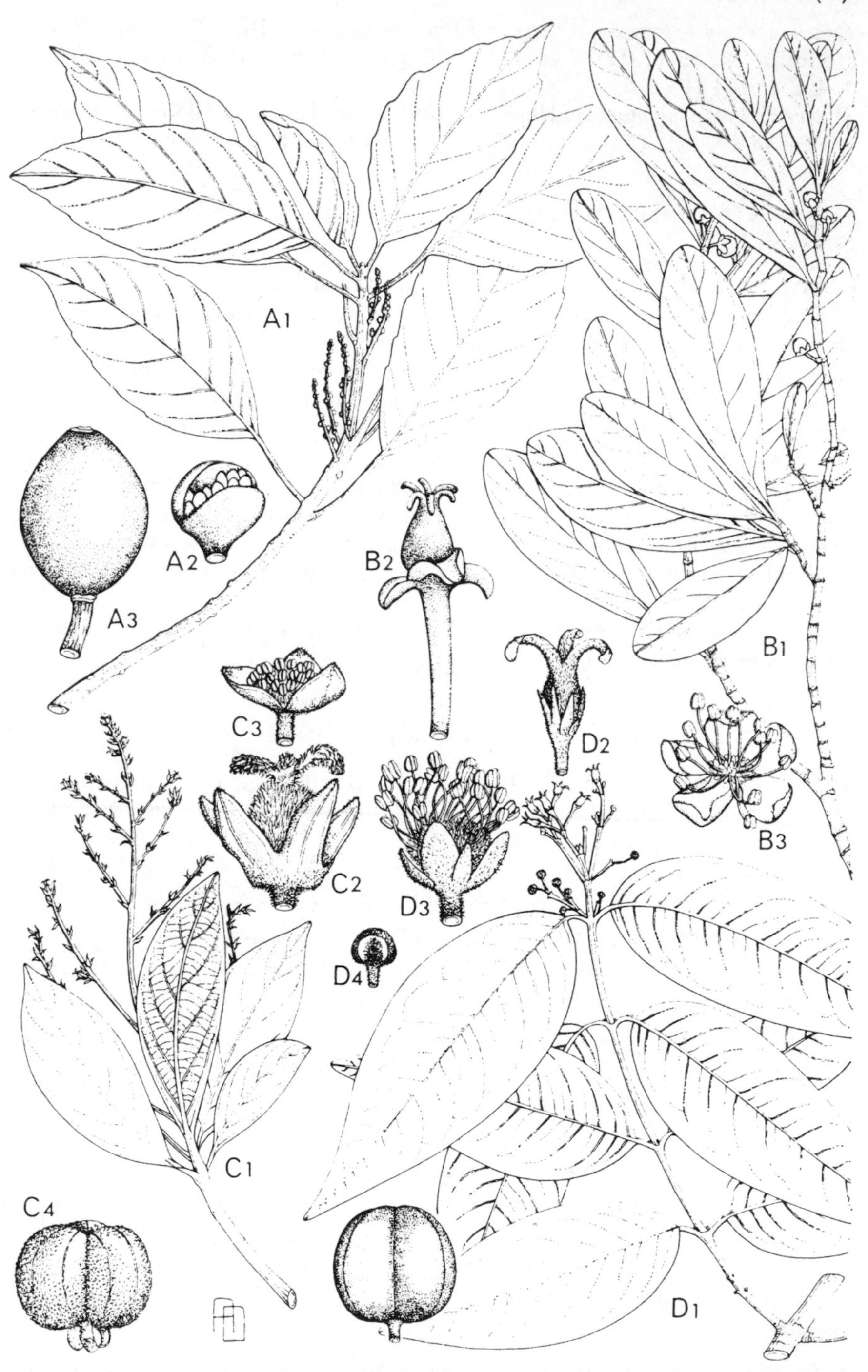

FIG. 6. *Pimelodendron amboinicum:* **A1** habit × $\frac{2}{5}$, from *Hyland* 7808; **A2** male flower × $2\frac{2}{3}$, from *Hyland* 7808; **A3** fruit × $2\frac{2}{3}$, from *Gafui* BSIP 10815. *Suregada glomerulata:* **B1** habit × $\frac{2}{5}$, from *Hyland* 3118; **B2** female flower × 4, from *Brass* 8219; **B3** male flower × 4, from *Brass* 8219A. *Rockinghamia brevipes:* **C1** habit × $\frac{2}{5}$; **C2** female flower × 4 & **C3** male flower × 2, all from *Hyland* 7778; **C4** fruit × 1, from *Hyland* 7779. *Whyanbeelia terrae-reginae:* **D1** habit × $\frac{2}{5}$, from *Hyland* 8300; **D2** female flower × 2, from *Hyland* 3052; **D3** male flower × 4, from *Hyland* 7744; **D4** bud × $2\frac{2}{3}$, from *Hyland* 8360; **D5** fruit × $\frac{2}{3}$, from *Hyland* 3052.

Rockinghamia angustifolia (*Benth.*) *Airy Shaw* in K.B. 20: 29 (1966); Francis: 435 (1970); Hyland: 13 (1971). Type: Q., Rockingham Bay, *Dallachy* (K).

Mallotus angustifolius Benth.: 141 (1873); Bailey, Syn. Queensl. Fl.: 478 (1883) & Queensl. Fl. 5: 1448 (1902) & Compreh. Cat. Queensl. Pl.: 479 (1913); Pax & Hoffm. vii: 204 (1914); Domin: 889 [335] (1927); C. T. White in Contrib. Arn. Arb. 4: 55 (1933); Francis, *op. cit.* ed. 2: 435 (1951).

Q (CK, NK, SK)—Endemic.

Tree to 20 m high, locally common in rain-forest or mesophyll vine forest on red clay soil up to 1200 m.

The pseudo-verticillate arrangement of the leaves, the mixed male + female inflorescences, and the presence of small glands amongst the stamens in the male flowers, immediately exclude this plant from the genus *Mallotus*. It differs from *R. brevipes* in the somewhat thinner leaves, with conspicuously reticulate venation and generally longer and slenderer petioles; in the male flower the stamens are longer and the pistillode is greatly reduced or lacking, whilst in the female the style-branches are more slender and bifid. The disk-glands or processes among the stamens are long-pilose, and may be subglobose or shortly clavate or variously confluent.

Rockinghamia brevipes *Airy Shaw* in K.B. 31: 389 (1976). Type: Q., Mt Bellenden-Ker, 1974, *Hyland* 7778 (K).

Q (CK)—Endemic.

Small tree of 5–7 m, in dwarfed rain-forest at 680–1500 m. Fig. 6C.

Closely related to *R. angustifolia*, but clearly distinct in its generally shorter petioles and more coriaceous leaves, with slightly fewer nerves and less conspicuous nervation in general. Less obvious distinctions are found in the shorter stamens and conspicuous pistillode of the male flowers, and in the thicker, simple, conspicuously papillose style-branches of the females.

Confined to the Bellenden-Ker/Bartle-Frere massif, mostly at elevations of 1300–1500 m, but occasionally down to 680 m.

Sauropus *Bl.*

(*Synostemon* F. Muell.)

Shrubs (sometimes scrambling) or shrublets, or herbs, sometimes from a woody base. Leaves alternate, entire, membranaceous or chartaceous, penninerved (rarely triplinerved), varied in size and shape, very shortly petiolate; stipules mostly small, triangular-subulate. Flowers monoecious or apparently dioecious, mostly fasciculate, axillary, or sometimes borne on special, densely bracteate, short or long brachyblasts, simulating racemose inflorescences, which may be cauliflorous from the old wood. Male flower: tepals 3 + 3, free and erect or spreading or variously connate, and then either forming a tube, or frequently flattened into an entire or acutely 6–12-lobed disk, the distal half of each tepal being sharply inflexed, so that the six apices point inward and fit closely around the androecium. (These apices have frequently been erroneously referred to as lobes of a disk.) Disk 0! Stamens 3, opposite the outer tepals, the filaments occasionally free, or usually connate into a short or very short column, the anthers borne either horizontally at the corners of the common triangular connective and facing downwards, or vertically at the apices of the filaments (when these are free), or often connate into an oblong-cylindric body. Female flower: tepals 3 + 3, biseriate, mostly ± turbinately connate below, often much larger than those of the male flower and accrescent in fruit, but sometimes very much smaller. Disk 0. Ovary 3-locular, sometimes apically truncate or excavated, loculi biovulate; styles either short, broad and entire or bifid, or deeply bifid with linear coiled or recurved segments, occasionally connate below into a conspicuous column. Capsule depressed-globose or ovoid, ± crustaceous, sometimes glossy, inconspicuously lobed, occasionally ± fleshy and berry-like; seeds triquetrous, exarillate, often dorsally rugose.

In *K.B.* 26: 343 (1971); *K.B. Add. Ser.* IV 199 (1975) & VIII: 205 (1980), I pointed out that the genus *Synostemon* F. Muell. (1858) is too close to *Sauropus* Bl., and recent more intensive study has convinced me that the two groups cannot be maintained as distinct taxa. Their bifocal development in Southeast Asia and Australia is curious and without an obvious parallel. It does not seem possible to utilize the subgenera and sections proposed by Mueller-Arg. in DC., Prodr. 15(2): 240–243 (1866) and by Pax & Hoffmann in Engl., Pflanzenr. IV. 147. xv: 216–226 (1922), in order to systematize the genus as a whole, including the Australian species. The so-called section (or subgenus) *Hemisauropus* Muell. Arg. (cf. Kew Bull. 23: 55 (1969)) appears to be unrepresented in Australia, and is in any case doubtfully tenable as a natural group, since the distinctive floral character seems to be uncorrelated with vegetative or other features. The majority of the Australian species of *Synostemon* would perhaps be included in *Sauropus* Sect. 4 *Schizanthi* Pax & Hoffm., but (like the Asiatic species) they seem to exhibit such a wide range of uncorrelated characters that any attempt to arrange them in formal groups would seem premature and unprofitable.

Apart from *Sauropus brunonis* (p. 671 below), which shows some evidence of affinity with *S. calcareus* Hend., of Malaya, the Australian species of *Sauropus* seem to show little or no obvious connection with the species of Southeast Asia, thus indicating a probable long separation of the two groups. It is curious that the most widespread member of the genus, *S. bacciformis* (L.) Airy Shaw (p. 685 below), has not yet been found on the coasts of Australia. It should be noted, however, that *S. macranthus* Hassk. extends from northeastern India through Malesia to northern Queensland.

1. Leaves large, 6–18 × 3–6 cm; stipules large, pale-margined; capsule large, to 2 cm diam.; pedicel up to 5 cm long . . . **S. macranthus**
1. Leaves small, less than 6 × 3 cm; stipules small, not pale-margined; capsule smaller, pedicel shorter:
    2. Leaves linear or narrowly oblong or oblanceolate, much longer than broad:
        3. Leaves hispidulous or papillose:
            4. Leaves linear . . . . **S. glaucus** var. **glaucus**
            4. Leaves elliptic, acute or mucronate **S. elachophyllus** var. **elachophyllus**
            4. Leaves cuneate-oblong . . . . **S. hirtellus**

3\. Leaves glabrous:
5\. Leaves acute at apex:
6\. Leaves linear or oblanceolate:
7\. Plant leafy at time of flowering . . . . **S. glaucus** var. **glaber**
7\. Plant almost leafless at time of flowering:
8\. Flowers minute, subsessile . . . . **S. lissocarpus**
8\. Flowers larger, pedicel 5–10 mm long . . **S. ramosissimus**
6\. Leaves elliptic . . . . **S. elachophyllus**
9\. Leaves 2–3 mm broad . . . . var. **glaber**
9\. Leaves up to 5 mm broad . . . . var. **latior**
5\. Leaves obtuse or rounded at apex:
10\. Leaves oblong, rounded at base:
11\. Leaves very stiff, glossy and grey when dry . . . **S. latzii**
11\. Leaves less stiff, not or scarcely glossy, greenish or ochreous or brownish and dull when dry:
12\. Plant ± bushy, 10–15 cm high . . . **S. trachyspermus**
12\. Plant erect, to 30 cm high . . . . **S. hubbardii**
10\. Leaves cuneate-obovate:
13\. Plant pinkish when dry; pedicels long, capillary . **S. albiflorus**
13\. Plant green when dry; pedicels short:
14\. Stipules obsolete; male perianth infundibular . **S. thesioides**
14\. Stipules subulate, 1–1·5 mm long; male perianth unknown
S. podenzanae (♂ plant)
2\. Leaves ovate, obovate, elliptic or suborbicular:
15\. Leaves glabrous:
16\. Leaves broadly elliptic:
17\. Stems erect; leaves very smooth . . . . . **S. ochrophyllus**
17\. Stems weak, ± prostrate; leaves minutely papillose beneath
S. podenzanae (♀ plant)
16\. Leaves narrowly elliptic . . . **S. elachophyllus** var. **latior**
15\. Leaves hispidulous or pubescent:
18\. Leaves deeply cordate, broadly ovate-orbicular, venose
S. ditassoides
18\. Leaves not cordate:
19\. Leaves very small, cuneate-obovate, hispidulous, under 5 mm long, fascicled . . . . **S. rigens**
19\. Leaves over 5 mm long:
20\. Leaves and stems softly pubescent; stipules black, conspicuous
S. rigidulus
20\. Leaves shortly hispidulous:
21\. Leaves with a prominent black apical gland . **S. crassifolius**
21\. Leaves without an apical gland:
22\. Male perianth tubular, pallid when dry . . . . . **S. huntii**
22\. Male perianth rotate, minute, black when dry . **S. brunonis**

Sauropus albiflorus (*F. Muell. ex Muell. Arg.*) *Airy Shaw* comb. nov.

Phyllanthus albiflorus F. Muell. ex Muell. Arg. in Linnaea 34: 70 (1865) & in DC., Prodr. 15(2): 326 (1866); Benth., Fl. Austr. 6: 100 (1873). Syntypes: Q., Edgecombe Bay & Moreton Bay, *F. Mueller* (G–DC).

Synostemon albiflorus (F. Muell. ex Muell. Arg.) Airy Shaw in K.B. 33: 37 (1978), *in adnot.*

Small glabrous shrub to 35(–120) cm tall, with numerous short slender often fascicled leafy brachyblasts 2–4 cm long. Leaves cuneate-obovate, 5–17 × 3–5 mm, rounded to subacute at apex, membranaceous, very smooth, often glaucescent beneath; midrib prominulous beneath, other nerves immersed, invisible; petiole 1 mm long; stipules and perulae subulate, 1–2 mm long, acute, castaneous. Flowers dioecious, borne on elongate capillary pedicels, males 5–10 mm, females up to 15(–20) mm long in fruit. Males: tepals obovate, subequal, 1·5 mm long, rounded at apex, glabrous; anthers connate in an oblong-cylindric mass nearly 1·5 mm long. Females: tepals obovate-spathulate, 2 mm long, acute; ovary globose, smooth, glabrous; styles 3(–4), free, erect, stout, stigmas shortly bilobed or reniform, horizontal. Capsule globose, 5–6 mm diam., smooth, glabrous; mature seeds not seen.

subsp. **albiflorus:** leaves up to 17 × 5 mm; capsule 5–6 mm diam.

Q (CK, NK, PC, BT, WB, MO)—Endemic.

Shrub at edge of creek in narrow belt of rain-forest in partial shade on gravelly soil, or very abundant among river-bed boulders, at unknown altitudes. Fig. 7B.

subsp. **microcladus** (*Muell. Arg.*) *Airy Shaw* stat. nov. Leaves less than half as large and capsule half as large as in subsp. *albiflorus*; styles deflexed, horizontal, with subulate lobes; ripe seeds quite smooth.

Phyllanthus microcladus Muell. Arg. α *puberulus* Muell. Arg. in Linnaea 34: 71 (1865) & in DC. Prodr. 15(2): 369 (1866); Benth., Fl. Austr. 6: 106 (1873). Type: N.S.W., Clarence River, *Beckler* (G–DC).

P. pusillifolius S. Moore in Journ. Linn. Soc., Bot. 45: 216 (1920). Type: Q., Broad Sound, *R. Brown* distrib. no. 3601 (BM).

Densely branched shrub up to 120 cm high; leaves green; in partially cleared 'scrub' (semi-rainforest) on brown loam, up to 100 m, or in depauperate rain-forest remnant on basalt hill, up to 460 m.

Q (PC, BT, MO); **NSW**—Endemic.

Sauropus albiflorus differs from all other species in its long filiform pedicels and thin cuneate-spathulate leaves.

Sauropus brunonis (*S. Moore*) *Airy Shaw* comb. nov.

Phyllanthus brunonis S. Moore in Journ. Linn. Soc., Bot. 45: 213 (1920). Type: N.T., Arnhem Bay, 1802–5, *R. Brown* distrib. no. 3608 (holotype BM; isotype K).

var. **brunonis:**

Prostrate herb from a perennial woody rootstock; branchlets slender, striate, hispid; leaves elliptic to lanceolate-elliptic, 1–3 × 0·5–1 cm, base mostly acute, apex acute to rounded-apiculate, margins ciliate-hispidulous and conspicuously reflexed, stiffly chartaceous, strongly elevate-reticulate-venulose on both surfaces, glaucous beneath, slightly glossy above, green or brownish when dry; petiole about 1 mm long; stipules minute, subulate, spreading, pilosulous, persistent. Flowers dioecious, solitary or few together, axillary, black when dry, on slender pilose or capillary glabrous pedicels 5–9 mm long. Males minute, 1–1·5 mm diam., tepals spathulate, truncate-obtuse; stamens connate in a short column, the connectives connate and expanded into a flat or shortly infundibular shallowly 6-lobed pileiform head, the minute anthers borne on the lower surface. Female flowers 3–4 mm diam., the tepals cuneate-obovate, subobtuse, setulose; ovary turbinate, flat-topped, glabrous, 1·25 mm diam., styles deeply bifid, with subulate arms spreading horizontally. Capsule globose-ovoid, 5–6 mm diam., smooth, glabrous. Seeds triquetrous, 5 × 3 mm, shallowly nodulose dorsally and laterally, excavated at the inner angle, dark brown when ripe.

NT (DG)—Endemic.

Herb to 7·5 cm high, with prostrate stems to 60 cm long, from perennial rootstock, in sandy loam in eucalypt forest. Fig. 7D.

var. **ovatus** *Airy Shaw* var. nov., foliis ovato-orbicularibus usque 13 mm latis, tepalis femineis late spatulato-ellipticis apiculatis.

NORTHERN TERRITORY. Near Nourlangie Safari Camp, 130 km NNE of Pine Creek Township, occasional in red sandy soil on gentle slopes, with *Eucalyptus miniata* and mixed scrub, 22 March 1965, *Lazarides & Adams* 293 (holotype K):—Prostrate perennial 30–60 cm long, 7·5 cm high, with discolorous leaves and greenish-yellow flowers; fruit carried on underside of stems. 60 km NE of Pine Creek, on granitic sands, 17 March 1971, *Dunlop & Byrnes* 2117:—Perennial rootstock—suffruticose herb.

In the leaf venation and the minute setulose flowers, drying black, *S. brunonis* shows evidence of affinity with *S. calcareus* Hend., of Malaya, and *S. amabilis* Airy Shaw, of West-Central and northeastern Thailand. In K.B. 23: 50–52 (1969), I proposed for *S. amabilis* and *S. villosus* (Blanco) Merr. the special section *Glochidioidei*, with *S. villosus* as the type, at the same time indicating that *S. calcareus* differed too much to be included in the same section. The same is true of *S. brunonis* (and the related *S. ditassoides* (Muell. Arg.) Airy Shaw): there are too many points of difference to warrant inclusion in the same group.

Sauropus crassifolius (*Muell. Arg.*) *Airy Shaw* comb. nov.

Phyllanthus crassifolius Muell. Arg. in Flora 47: 513 (1864) & in DC., Prodr. 15(2): 325 (1866); Benth., Fl. Austr. 6: 100 (1873). Type: W.A.: Sharks Bay, *Milne* (G–DC).

Glochidion ecrassifolium (sphalm.) C. A. Gardn. ex Beard, W. Austr. Pl.: 58 (1967).

A rigid densely twiggy shrublet, occasionally up to 120 cm in height; branches terete, woody, minutely papillose-puberulous; lateral branchlets

3–10 cm long, often fascicled, ascending or arcuate-decurved, densely leafy; leaves small, distichous, broadly obovate-orbicular, 5–7 × 3–6 mm, rigidly coriaceous, densely puncticulate-scabrous, tipped with a prominent black gland, nerves few, inconspicuous, petiole almost obsolete, stipules and perulae small, ovate, acute, black when dry, forming conspicuous nodules on the branchlets after the leaves have fallen. Flowers in small subsessile mixed axillary fascicles. Male flower: tepals ovate, dorsally keeled, acute, glabrous, slightly hooded, the outer 1·5 mm, the inner 2 mm long; stamens 3, the anthers connate in an ovoid mass, 1 mm long, connectives short, distinct, obtuse. Female flower: tepals as male flower, but outer tepals minutely pilosulous, with a small black apical gland; ovary depressed-globose, glabrous; styles 3, sigmoid, patulous below, connivent and thickened apically, with shortly bifid stigmas. Capsule not seen.

WA (ER—Austin)—Endemic.

No field information. Fig. 7C.

The conspicuous black apical gland of the leaves distinguishes *S. crassifolius* from all other species.

Sauropus ditassoides (*Muell. Arg.*) *Airy Shaw* comb. nov.

Phyllanthus ditassoides Muell. Arg. in Flora 47: 487 (1864) & in DC., Prodr. 15(2): 326 (1866); Benth., Fl. Austr. 6: 97 (1873). Type: N.T.: In Nova Hollandia septentrionali, *Armstrong* (K).

Undershrub with prostrate slender angled flexuous subsimple stems, shortly branched above, shortly spreadingly whitish-pilosulous, up to 35 cm long, from a thick woody subterranean rootstock. Leaves ovate or almost oblong when young, becoming broader and suborbicular later, rounded or cordate at base, acute or rounded-apiculate at apex, up to 3·5 × 2·7 cm, margin narrowly reflexed, chartaceous, plumbeous when dry and slightly purplish beneath, sparsely scaberulous-pilose above and on the margins, almost glabrous beneath except on the nerves, subsessile; stipules minute, subulate, brownish, quickly caducous. Flowers apparently dioecious (or males and females borne on separate stems of the same plant?), males in small fascicles, females mostly solitary. Males: tepals ± spathulate, infundibular-connate below, spreading above, obtuse or subacute, greenish-yellow in life, glabrous or densely hispidulous; anthers connate in an oblong-cylindric mass 2–3 mm long, connectives rounded. Females: tepals as in males; ovary depressed-globose, puberulous; styles very short, with bifid U-shaped stigmatic arms. Capsule not seen.

NT (DG)—Endemic.

Prostrate perennial-rooted herb to 35 cm, with pale green rather discolorous leaves, on sandy soil in woodland or occasional on gentle slopes in gritty whitish grey sandy soil with laterite pebbles, associated with *Eucalyptus tetrodonta*, *E. miniata*, dense *Livistona humilis*, annual *Sorghum*, and short annual grasses and sedges.

Sauropus elachophyllus (*F. Muell. ex Benth.*) *Airy Shaw* comb. nov.

Phyllanthus elachophyllus F. Muell. ex Benth., Fl. Austr. 6: 101 (1873). Syntypes: Q., Newcastle Range, *F. Mueller* (lectotype K, present designation); Einasleigh river, *Daintree* (MEL).

var. **elachophyllus:** stems and leaves hispidulous-puberulous.

An erect woody shrublet, to 100 cm tall, with numerous slender, angled, often fascicled, ± virgate, minutely puberulous branches. Leaves elliptic-oblong, distichous, 2–6 × 2–3 mm, subsessile, acute, rounded at base, stiffly chartaceous, minutely softly papillose-puberulous, or sometimes hispidulous, midrib slender, slightly prominent below, lateral nerves immersed; stipules and perulae minute, subulate, fuscous, crowded, fimbriate, forming small dark nodules conspicuous along the branchlets after the leaves have fallen, apparently representing greatly contracted racemose inflorescences. Flowers in the axils of the perulae, very shortly pedicelled, pendulous on the underside of the branchlets. Males minute, rotate, less than 1·5 mm diam., tepals ± oblong, obtuse, dorsally very minutely pilosulous, glabrous within, pedicel short and very slender; anthers connate in an oblong-cylindric mass, connectives short, obtuse. Females larger, with thicker perianth, ± infundibular-spreading, lobes acute; ovary glabrous; styles bifid, erect or incurved. Capsule (from fragments) ± 5 mm long, smooth, glabrous; seeds smooth.

Q (CK, NK)—Endemic.

var. **glaber** *Airy Shaw* var. nov., omnino glaber, foliis angustis ut in var. *elachophyllo*.

QUEENSLAND. Cook District: Cape York Peninsula, 1873, *W. Hann* 220 (K). Telegraph Line, between Archer River and Coen, common in sandy messmate-bloodwood savanna forest, 150 m, 1 Aug. 1948, *Brass* 19770 (♀) (holotype K), 19771 (♂):—Shrub 60–100 cm tall; leaves grayish, fleshy. Isabella Falls, 43 km NW of Cooktown, open forest, 18 June 1972, *Wrigley & Telford* NQ 1370 (seed 722439).

var. **latior** *Airy Shaw* var. nov., omnino glaber, foliis oblongo-ellipticis usque 5 mm latis.

NORTHERN TERRITORY. 10 miles [14 km] S of Yaimanyi Creek, 12° 53′ S, 134° 34′ E, among sandstone rocks, 24 June 1972, *Byrnes* 2693. Oenpelli, in mixed open forest at foot of sandstone hill, 18 Oct. 1948, *Specht* 1212 A & B:—Bush 60 cm tall, 45 cm diam., flowers green.

QUEENSLAND. Cook Distr.: 75 km from Laura on Coen Road, 15° 10′ S, 143° 50′ E, open forest, 60 m, 24 Sept. 1976, *Hyland* 9043:—Shrub 1 m tall (holotype K).

In leaf-shape var. *latior* approaches the smaller-leaved forms of *S. ochrophyllus*, but it exhibits the rough texture and dark fuscous-grey colour of *S. elachophyllus* in drying, showing none of the smooth glaucous-ochraceous appearance of *ochrophyllus*. Further material might indicate that it deserves specific rank.

Sauropus glaucus (*F. Muell.*) *Airy Shaw* comb. nov.

Synostemon glaucus F. Muell., Fragm. 1: 33 (1858). Syntypes: N.T., Arnhem land; MacAdam Range, Point Pearce & Providence Hill, *F. Mueller* (MEL).

Phyllanthus stenocladus Muell. Arg. in Flora 47: 536 (1864) & in DC., Prodr. 15(2): 327 (1866). Type: N.T., Port Essington, [*Armstrong*] 503 (K).

Phyllanthus adami Muell. Arg. in DC., Prodr. 15(2): 327 (1866); Benth., Fl. Austr. 6: 97 (1873). Type: N.T., McAdam Range, *F. Mueller* (G-DC).

Glochidion adamii (Muell. Arg.) C. A. Gardner ex Beard, W. Austr. Pl.: 58 (1967).

var. **glaucus:** Stems minutely hispidulous; leaves hispidulous.

Shrublet to 30 cm high, stems erect from a woody rootstock, much branched, virgate, strongly angled, blackish when dry, minutely hispidulous; leaves linear or narrowly oblanceolate, sessile, apex acute or rounded and mucronate, 5–14(–25) × 1–3(–10) mm, minutely pustulate and hispidulous, sometimes glaucescent beneath, margins subrevolute. Flowers dioecious, solitary, axillary, cernuous, black when dry, very shortly (1 mm) pedicelled. Males: tepals obovate, subequal, 2–3 mm long, dorsally glabrous or minutely hispidulous; anthers connate in an oblong mass occupying the upper two thirds of the staminal column. Females: tepals linear, 3–4 mm long, acute, glabrous, at first erect, later spreading; ovary with 3 deeply bifid styles with linear segments 2 mm long; pedicel up to 7 mm long. Capsule ovoid, 6–7 mm long and wide, shortly apiculate, smooth, glabrous, with a dull surface. Seeds narrowly triquetrous, 8–9 × 2 mm, minutely asperulous.

NT (DG)—Endemic.

Erect from perennial base, pink stems, rare in sandy soil in open *Eucalyptus tetrodonta* forest (Elcho I.).

var. **glaber** *Airy Shaw* var. nov. Tota planta omnino glaber. Typus: N.T., Doingi Airstrip, *J. Must* 1050 (holotypus K).

Phyllanthus ochrophyllus sec. Dunlop, Latz & Maconochie in N. Terr. Bot. Bull. 1: 22 (1976), p.p.

NT (DG)—Endemic.

Mixed with var. *glaucus* in *Brown's* (distrib. no. 3608) and *Cunningham's* (Port Keats, no. 475) collections. Also: 'Subtrop. N. Holl.', *Major Mitchell* s.n.; Port Essington, Aug. 1839, *Armstrong* 501; Port Darwin, March 1870, *F. Schultz* 297, 460. In *Eucalyptus tetrodonta* open forest on truncated lateritic podsol, Gove, 12° 15′ S, 136° 43′ E, 7 Sept. 1948, *Specht* 1028:—Bush 30 cm tall and wide, flowers pale yellow-green. Herb to 15 cm high. 18 miles [27 km] W South Alligator R., Jim Jim road, in eucalypt woodland 13 Nov. 1971, *Must* 887. U.D.P. Mine Area, sandstone slope, growth after fire, 17 March 1971, *Dunlop & Byrnes* 2124:—Small herb from perennial rootstock. 66 km from Pine Creek towards U.D.P. Falls, July 1973, *Gittins* 2686:—Shrub 0·5 m high with white flower; female has 6 much curved styles; male is dry and brittle with a few short stamens. 16 m [24 km] SW

Maningrida, herb to 30 cm high, regrowth after fire, Eucalypt woodland, lateric soil, 3 Sept. 1971 *Must* 783. Doingi Airstrip, 12° 54′ S, 135° 28′ E, lateric soil—disturbed site, 23 June 1972, *Must* 1050 (holotype K). McMillans Road area, Darwin, 12° 24′ S, 130° 54′ E, open forest; lateritic soil, 20 Sept. 1977, *Parker* 1100:—Perennial herb; flowers greenish yellow; 10 cm. Bamboo Creek, 22 m W Batchelor, sandy hill, creekside, 20 Oct. 1972, *McKean* 13729:—Herb 35 cm high, white flower. Elcho Island, 11° 59′ S, 135° 43′ E, rare in sandy soil, open *Eucalyptus tetrodonta* forest, 2 July 1975 *Latz* 6070:—Erect, perennial base, pink stems.

Sauropus hirtellus (*Muell. Arg.*) *Airy Shaw* comb. nov.

Synostemon hirtellus F. Muell., Fragm. 3: 89 (1862). Type: Q., (PC): In Nova Hollandia orientali tropica prope Walloon, *Bowman* (MEL).
Phyllanthus hirtellus (F. Muell.) Muell. Arg. in DC., Prodr. 15(2): 326 (1866); Benth., Fl. Austr. 6: 98 (1873).
Glochidion hirtellum (F. Muell.) Hj. Eichl., Suppl. Black's Fl. S. Austr.: 210 (1965).

Slender subshrub with erect terete branched papillose or minutely hispidulous stems to 35 cm tall, 0·5–2 mm thick. Leaves distant, very narrowly spathulate-oblanceolate, 4–15 × 1–2 mm, rounded and mucronate at apex, papillose, straight-sided, gradually narrowed to the cuneate base, erect-ascending, practically sessile; stipules minute, subulate. Flowers solitary, axillary. Males: narrowly tubular-infundibular, 4 mm long, dark brown when dry, with very short rounded tepals 1 mm diam., staminal column 4 mm long, the linear anthers occupying the upper half, connectives very shortly produced, obtuse. Female flowers not seen; tepals (in fruiting stage) suborbicular, 1 mm diam., persistent. Capsule subglobose, almost 1 cm diam., smooth, pallid, glabrous, seeds triquetrous, 6 × 3 mm, pale grey, finely roughened and very shallowly ridged on the back, inner angle excavated.

Q (PC)—Endemic.

No field information.

Only known from the type and *McGillivray* (voyage of Rattlesnake) B105 (K) from Port Curtis, Nov. 1847.

Sauropus hubbardii *Airy Shaw* sp. nov., *S. trachyspermo* (F. Muell.) Airy Shaw affinis, sed habitu elatiore erecto usque 40 cm alto, foliis 2–3 cm longis, tepalis apice obtusis vel rotundatis, stylis brevissimis sed alte bipartitis recedit. Typus: Q. (BK), *Hubbard & Winders* 7295 (holotypus K).

Suffrutex herbaceus e caudice lignoso exortus, glaberrimus, caulibus erectis usque 40 cm altis basi 2–3 mm crassis teretibus tenuiter sulcatulis superne gracillimis compressis vel angulatis, ramis adscendentibus. *Folia* anguste vel angustissime oblonga, juniora etiam linearia, usque 25 × 5 mm vel 31 × 3 mm, basi anguste usque late cuneata vel anguste rotundata, apice rotundato usque acuto semper minute mucronulato, margine integro anguste incrassato

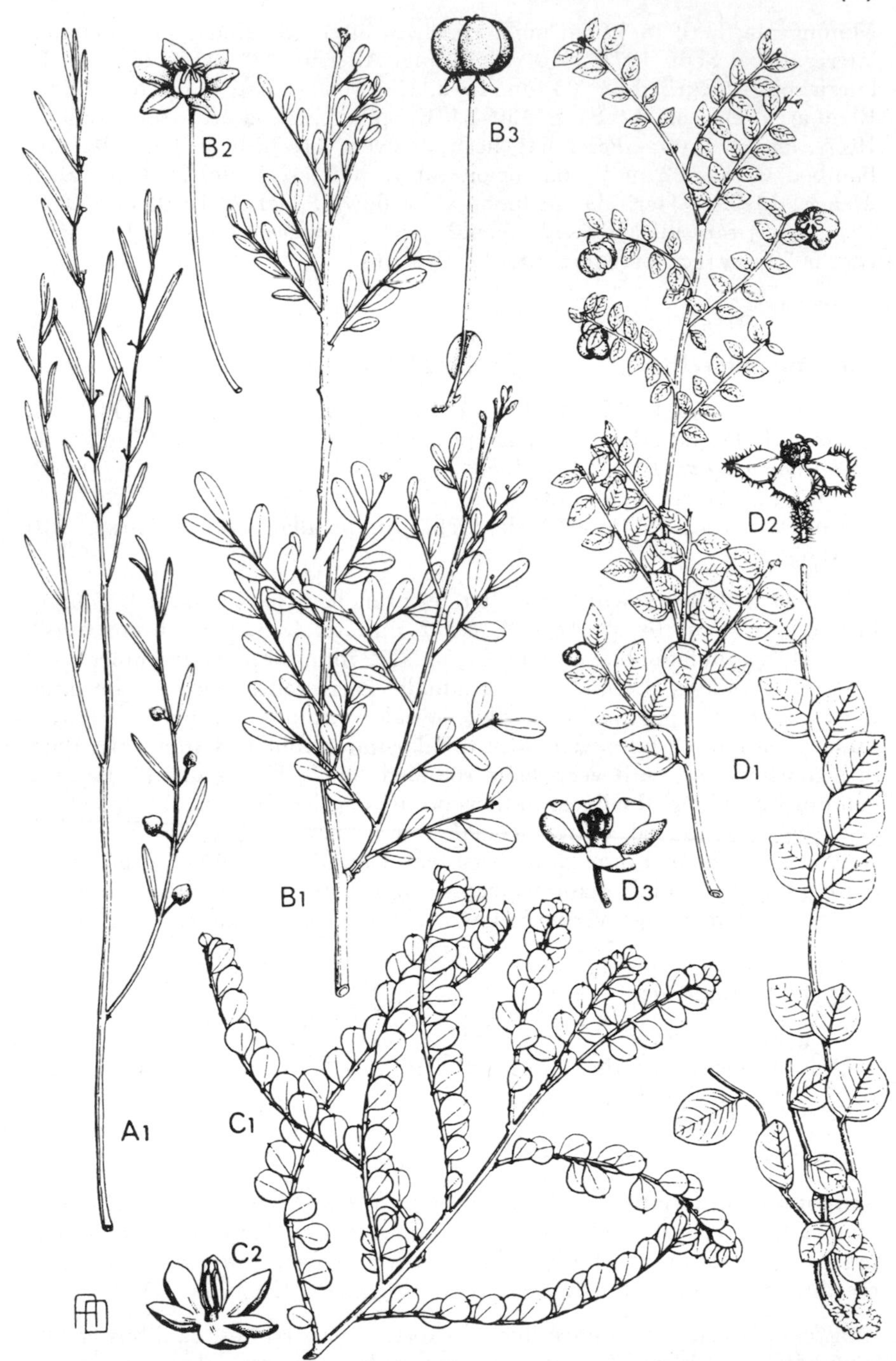

FIG. 7. *Sauropus hubbardii:* **A1** habit × ⅔, from *Winders* 7295. *Sauropus albiflorus* subsp. *albiflorus:* **B1** habit × ⅔, from *Clemens* 1947; **B2** female flower × 4, from *Hubbard* 3210; **B3** fruit × 2⅔, from *Hubbard* 3210. *Sauropus crassifolius:* **C1** habit × ⅔, from Herb. Hooker 1867; **C2** male flower × 4, from Herb. Hooker 1867. *S. brunonis* var. *ovatus:* **D1** habit × ⅔, from *Lazarides & Adams* 293; **D2** female flower × 6, from *Lazarides, Dunlop & Byrnes* 2117; **D3** male flower × 8, from *Lazarides & Adams* 293.

et reflexo, herbacea, laevia, siccitate viridula; costa gracilis, subtus prominula, supra fere plana; petiolus vix 1 mm longus; stipulae anguste subulatae, 1 mm longae, acutissimae, ± erectae, persistentes. *Flores* monoeci, axillares, solitarii vel rarius bini. *Flos* ♂ pedicello 1 mm longo suffultus; tepala 3 + 3, ovato-orbicularia, vix 1 mm longa et lata, exteriora paullo majora, basi breviter connata, apice rotundata; stamina 3, filamentis connatis, antheris in massam breviter cylindricam connatis, thecis apice distinctis longitudinaliter dehiscentibus. *Flos* ♀ pedicello 3 mm longo suffultus; tepala 3 + 3, oblongo-obovata, fere 2 mm longa, basi breviter connata, apice obtusa; ovarium breviter cylindricum, 1 mm longum, sulcis 3 notatum, apice truncatum; styli 3, brevissimi, suberecti, usque ad basin bifidi, in vertice ovarii late sejuncti. *Capsula* ovoidea, 4–6 mm longa, apice subacuto, leviter trisulcata, laevis, pallide glaucescens. *Semina* trigona, 3–5 × 2 mm, hilo excavato, dorso convexo tenuiter pulchre lamellato-muriculato, matura saturate brunnea. Fig. 7A.

QUEENSLAND. Burke Distr.: Nonda, between Hughenden and Cloncurry, in mixed grassland in heavy dark brown soil, 160 m, 6 Feb. 1931, *Hubbard & Winders* 7295 (holotype K):—Erect herb with dull green leaves.

The relatively tall erect habit, considerably longer leaves, obtuse or rounded instead of sharply acute tepals, and completely bipartite (though still very short) styles, distinguish this plant from the related *S. trachyspermus* (*Phyllanthus rhytidospermus* F. Muell. ex Muell. Arg.). The latter species seems not yet to be known from North Queensland (specimens seen from Warrego, Gregory South, Mitchell and Port Curtis). Nonda is a township some 70 miles [100 km] west of Richmond, on the Flinders Highway.

It gives me pleasure to name this new species after my former colleague, the late Dr C. E. Hubbard, one time Keeper of the Herbarium, Royal Botanic Gardens, Kew, and internationally known for his work on grasses. During his tour of Australia in 1931 he covered an immense amount of ground and amassed a large quantity of most valuable herbarium material.

Sauropus huntii (*Ewart & Davies*) *Airy Shaw* comb. nov.

Phyllanthus hunti Ewart & Davies, Fl. N. Terr.: 164, t. xvi (1917). Type: N.T. (CA): 130 miles N of N.T. Survey Camp IV., 2 July 1911, *G. F. Hill* 130 (MEL).

Shrub with long ascending or spreading branches to 40 cm long, terete, hoary-papillose throughout. Leaves broadly elliptic, distichous, 10–16 × 5–11 mm, rounded or very slightly cordate at base, rounded or slightly emarginate or occasionally mucronate at apex, firmly chartaceous, nerves few, patulous, prominulous beneath; petiole 1 mm long or less; stipules narrowly subulate, acute, 2 mm long, ± persistent. Flowers dioecious, in small axillary fascicles. Males cylindric-tubular, 5–6 mm long, 2 mm diam., papillose, pallid, with exceedingly small erect rounded-ovate tepals 1–1·5 mm diam.; filaments connate in a column 4 mm long, anthers narrow-elongate with linear thecae 2 mm long attached to the upper half of the column, connectives shortly produced as small erect ovate appendages; pedicel 2mm long. Female flowers: tepals free, narrowly spathulate, subequal, 4–5 × 1–2 mm, subacute, papillose; ovary ovoid or subglobose, minutely papillose, styles short, erect, free, with small bilobed stigma. Capsule 5–7 mm pedicelled, large, ovoid-globose, to 9 mm long, pale, smooth, minutely papillose, with persistent spreading tepals; seeds triquetrous, ochraceous, 7–8 × 2–3 mm, dorsally rugose-costate, laterally rugulose, inner angle excavated.

NT (CA)—Endemic.

The type has not been seen; the above description was drawn up from *Maconochie* 1693 & 1696 (NT), 13 miles NE Lake Surprise, Lander River, 20° 02′ S, 131° 58′ E, NNW of Alice Springs.

Sauropus latzii *Airy Shaw* sp. nov., *S. elachophyllo* (F. Muell.) Airy Shaw affinis, a quo foliis oblongis laevissimis nitidis sed siccitate rugosulis apice rotundatis marginibus incrassatis differt. Typus: N.T. (GD): *Latz* 3362 (N.T. 36884) (holotypus K; isotypi DNA, NT).

[*Phyllanthus ochrophyllus* sec. Dunlop, Latz & Maconochie in Northern Terr. Bot. Bull. 1: 22 (1976), p.p., *non* Benth.]

Suffrutex parvus, glaber, ramis gracilibus patentibus acute angulatis usque 25 cm longis. *Folia* oblonga, 7–15 × 3–5 mm, basi et apice rotundata, vel apice interdum acuto, rigide chartacea, laevissima sed siccitate rugulosa, viridi-grisea, margine reflexo leviter incrassato, costa infra elevata supra impressa, nervis immersis; petiolus fere obsoletus; stipulae et perulae minimae, acutae, fusco-brunneae. *Flores* dioeci (?), in fasciculis parvis axillaribus dispositi. *Masculi*: tepala spatulata, 3 mm longa, glabra, exteriora 1 mm lata, apice rotundata, interiora angustiora, obtusa; stamina in columnam crassam 3 mm longam connata, antheris linearibus 1·25 mm longis apici columnae affixis (apice ut videtur truncato), connectivis vix productis obtusis. *Flores feminei* et *fructus* non visi.

Northern Territory (DG). Wessel Islands [Elcho I.], 11° 11′ S, 136° 44′ E, rare in lateritic gravel, 1 Oct. 1972, *Latz* 3362 (NT 36884):—Small spreading subshrub; 'female plant, fruit globular'.

Close to *S. elachophyllus* var. *glaber*, but apparently distinct in its very smooth oblong leaves rounded at base and apex, the margins thickened and narrowly reflexed. Only the male plant has been seen; the collector appears to have mistaken male flower-buds for young capsules. The material is somewhat scrappy.

Sauropus lissocarpus (*S. Moore*) *Airy Shaw* comb. nov.

Phyllanthus lissocarpus S. Moore in Journ. Linn. Soc., Bot. 45: 215 (1920). Type: N.T. (DG), Groote Eylandt, *R. Brown* distrib. no. 3606 (holotype BM).

[*P. ochrophyllus* sec., Dunlop, Latz & Maconochie in N. Terr. Bot. Bull. 1: 22 (1976), p.p., *non* Benth.]

Many-stemmed virgate dioecious subshrub from a woody rootstock, often almost leafless at time of flowering, 30–40 cm high; stems angled, glabrous, much branched. Leaves sparse, linear or oblong, 0·5–2·3 cm × 1–3 mm, subsessile, rounded and apiculate at apex, stipules minute, subulate. Flowers very small, subsessile along the upper part of the leafless branches. Males 1–2 mm diam.; tepals obovate, obtuse, spreading, glabrous; filaments connate in a column, anthers quite free (*sec.* Moore, sed?). Females 2 mm diam., tepals obovate, similar; ovary turbinate, truncate above, glabrous; styles short, thick, bifid. Capsule and seeds not seen.

NT (DG); ? **Q** (CK)—Endemic.

Green almost leafless subshrub 20–30 cm high, with *Eucalyptus microtheca* and *Eulalia fulva* in swampy, heavy soil area, or infrequent on stable coastal dunes with *Triodia* and *Spinifex*.

The minute, almost rotate flowers, subsessile along the leafless branches, provide a ready distinguishing character for this species from other leafless taxa.

The reason for the application of the epithet *lissocarpus*, 'smooth fruit', to this species, when the fruit was unknown, is not immediately obvious.

Sauropus macranthus *Hassk.*, Retzia 1: 166 (1855) & Hort. Bog. s. Retziae ed. nova: 52 (1858); Muell. Arg.: 240 (1866); Backer & Bakh. f.: 471 (1963); Airy Shaw in K.B. 26: 336 (1972) & K.B. Add. Ser. IV: 193 (1975), *q.v.* for detailed synon. Type: Java, cult. in Hort. Bogor., 1854?, *Hasskarl* (BO).

S. spectabilis Miq., Fl. Ind. Bat., Suppl.: 446 (1861); Muell. Arg.: 240 (1866); Pax & Hoffm. xv: 219 (1922); Adelbert & Meeuse apud Backer in Blumea 5: 508 (1945). Type: Sumatra, in prov. Lampong, in m. Radjabassa, *Teijsmann* (BO).

Glochidion umbratile Maiden & Betche in Proc. Linn. Soc. NSW 30: 370 (1905); Domin: 873 [319] (1927), *in obs.* Type: Q., Atherton, 1901, *Betche* s.n. (NSW).

Q (CK)—NE India, S China, SE Asia, and scattered through Malesia to New Guinea.

Shrub to 2·5 m tall, in rain-forest at 750–800 m.

Entirely glabrous. Stem erect, with almost horizontally spreading branches, generally with an exceedingly narrow wing or keel decurrent from the base of each petiole. Leaves ovate, up to 18 × 6 cm in Malesia, not seen larger than 9 × 4 cm in Australian material, broadly cuneate at the base, gradually narrowed to the shortly acuminate apex, olive-green when dry, herbaceous to chartaceous, midrib and nerves elevated on both surfaces; petiole 3–7 mm long; stipules triangular, 2–5 mm long, acute, conspicuously pale-margined. Flowers blackish red, 1–2 together from distal axils, the males very small and short-pedicelled, the females larger on longer pedicels. Capsule large, 1·5–2·5 cm in diameter, fleshy, ± indehiscent, reddish, pedicel up to 5 cm long.

The plant bears an extraordinary superficial resemblance to *Phyllanthus sauropodoides* Airy Shaw. In Australia it appears to be almost confined to a small area of rain-forest to the south of Atherton.

Sauropus ochrophyllus (*Benth.*) *Airy Shaw* comb. nov.

Phyllanthus ochrophyllus Benth., Fl. Austr. 6: 99 (1873). Syntypes: N.T. (DG): Port Darwin, *Schultz* 428 (♂) & 489 (♀) (K).

An erect, glabrous, very smooth, ochreous-glaucescent perennial, the stems up to 40 cm long, simple and leafless below, 3–4-angled, with erect or ascending leafy lateral branches above. Leaves elliptic when young, 7–13 × 3–6 mm, very broadly orbicular-elliptic when mature, subsessile, narrowly rounded at base, obtuse to subacute at apex, margin flat, stiffly chartaceous, glabrous, 3-nerved at base, nerves almost immersed, stipules minutely subulate or obsolete. Flowers monoecious, solitary, axillary, pale yellow. Males: tepals narrow-oblong, erect, 3–4 mm long, acute and ± reflexed at apex; anthers elongate, connate in an oblong-cylindric body 2·5 mm long, connectives obtuse, slightly prominent. Females: tepals similar but slightly broader, less spreading at apex; ovary subglobose, 4–5 mm diam., smooth, glabrous; styles connate below in a column 2 mm long, erect, thickened and shortly bifid at the apex. Capsule globose, 8 mm long and diam., smooth, glabrous, remains of styles conspicuous at apex; seeds smooth.

NT (DG)—Endemic.

Only one other collection seen besides the type: Rum Jungle, 18 Oct. 1967, *W. F. Ridley* 78 (BR, K).

Sauropus podenzanae (*S. Moore*) *Airy Shaw* comb. nov.

Phyllanthus podenzanae S. Moore in Journ. Linn. Soc., Bot. 45: 214 (1920). Type: Q., (CK): Cooktown, *Podenzana s.n.* (BM).

Weak dioecious undershrub, almost glabrous, with slender much-branched angled stems to 30 cm long; leaves of male plant elliptic, herbaceous, greenish when dry, 7–12 × 2–4 mm, glabrous above, papillose-glaucescent beneath, obtuse or subacute at apex, margin reflexed; leaves of female plant obovate-orbicular, to 2 cm long and 1·7 cm wide, papillose-glaucescent beneath, nerves few, ascending, prominulous; petiole 1–2 mm long; stipules narrowly subulate, 1–2 mm long. Male flower not seen. Female flower solitary, axillary; tepals free, oblong-ovate, obtuse, membranaceous, glabrous; ovary glabrous, styles entire, revolute, slightly divergent.

Q (CK)—Endemic.

Only known on the banks of the Endeavour River near Cooktown.
'Perennial with dull pale green leaves, glaucous beneath; white flowers—dioecious, female with broad leaves'. On river levee in cleared open forest on sandy rise.

Sauropus ramosissimus (*F. Muell.*) *Airy Shaw* comb. nov.

Synostemon ramosissimus F. Muell., Fragm. 1: 33 (1858). Syntypes: Q.; In montibus saxosis ad flumina Suttor et MacKenzie River, Barrier Range (Mackenzie Range), *Becker* (G–DC); Darling River, *F. Mueller* (MEL).
Phyllanthus ramosissimus (F. Muell.) Muell. Arg. in Linnaea 34: 70 (1865) & in DC., Prodr. 15(2): 326 (1866).

Slender much branched subshrub to 30 cm or more high, stems erect, virgate, angled, glabrous, leafless or almost so. Leaves linear or cuneate-oblanceolate, 2–12 × 1–2 mm, glabrous, subacute, subsessile; stipules minute, subulate, reddish-brown. Flowers in very small axillary fascicles. Males very shortly pedicelled, tepals ovate, 1 mm long, anthers connate in an elongate oblong mass. Females 4–5 mm long; tepals linear, acute; pedicel 4–10 mm long; styles (not seen) free, thick, dilated and shortly bilobed. Capsule ovoid, 6–8 mm long; seeds strongly and irregularly transversely nodulose-rugose.

Q (LT, PC); **NSW**; **SA**—Endemic.

Both male and female flowers of this species are borne on slender pedicels, which may exceed 1 cm in the fruiting stage. The erect, very narrowly oblong tepals, especially of the males, confer a narrowly cylindric outline on the flower in the bud stage. The plant seems not to have been collected again since the time of Mueller; the latest collection at Kew is that of *R. Helms s.n.*, 25 May 1891, from Arco-eillinna Well in South Australia, distributed as '*Phyllanthus thesioides*'.

Sauropus rigens (*F. Muell.*) *Airy Shaw* comb. nov.

Synostemon rigens F. Muell., Fragm. 2: 153 (1858). Syntypes: N.S.W.: Upper Darling River, *Bowman*; Mutanic Range, *Beckler* (MEL).
Phyllanthus rigens (F. Muell.) Muell. Arg. in Flora 47: 513 (1864) & in DC. Prodr. 15(2): 325 (1866); Benth., Fl. Austr. 6: 99 (1873).
Heterocalymnantha minutifolia Domin in Biblioth. Bot. 22: 867, t. 32, figs. 1–9 (Heft 89: 313) (1927), **synon. nov.** Type: Q. (MI): 'Sandsteinhügel der Dividing Range bei Jericho, 1910', *Domin* s.n. (PR).
Glochidion rigens (F. Muell.) Hj. Eichler, Suppl. Black's Fl. S. Austr. ed. 2: 210 (1965).

Stiff twiggy shrublet, glabrous, to 100 cm high, branches terete, rigid, minutely hispidulous, bearing very small leaves fascicled on very short brachyblasts, shortly cuneate-obcordate, 2–6 mm long, to 3 mm wide, subsessile, emarginate at apex, minutely hispidulous, petiole almost obsolete; stipules and perulae minute, triangular-subulate, densely crowded, black-purple when dry, forming a conspicuous pulvinus at each brachyblast. Flowers monoecious, solitary at each brachyblast. Males: tepals connate, forming a narrowly infundibular perianth 3–4 mm long, clavate in bud, minutely papillose-hispidulous, lobes very small, ovate, less than 0·5 mm long, erect, acute; pedicel 1–2 mm long; anthers linear, almost 2 mm long, connate, occupying upper half of staminal column, connectives very shortly produced, obtuse. Females: tepals free or shortly connate, oblong-spathulate, almost 3 mm long, less than 1 mm broad, subobtuse, minutely papillose-hispidulous externally; ovary trigonous, 1 mm long, glabrous, passing into a stout trigonous style 1 mm long, bearing 3 small minutely bilobed capitate stigmas. Capsule (mature not seen) globose, 5–6 mm diam., smooth, glabrous, seeds triquetrous, strongly and sparsely reticulate-foveate when ripe; pedicel to 8 mm long, hispidulous.

NT (CA), **Q** (MI), **NSW** (FW), **SA**—Endemic.

Shrub to 1 m, new stems brown; rare in skeletal soil on side of sandstone hill, or in exceedingly dry bush margins near railroad, or scarce near old opal mines.

An unmistakable plant from its rigid twiggy habit and very small hispidulous spathulate-obcordate leaves in sessile fascicles arising from small cushions of minute black perulae. The plate given by Domin, *l.c.* (1927), gives an excellent idea of the plant.

Sauropus rigidulus (*F. Muell. ex Muell. Arg.*) *Airy Shaw* comb. nov.

Phyllanthus rigidulus F. Muell. ex Muell. Arg. in Linnaea 34: 72 (1865) & in DC., Prodr. 15(2): 370 (1866); Benth., Fl. Austr. 6: 99 (1873). Type: N.T. (DG): Gulf of Carpentaria, *F. Mueller* (G–DC).

Rigid erect shrub to 60 cm high, branches woody at the base, 5 mm thick, densely and finely tomentellous, branchlets erect or ascending, finely striate, leafy. Leaves broadly elliptic, 7–12 × 4–8 mm, rounded at base, rounded and minutely cuspidate or mucronate at apex, stiffly chartaceous, minutely tomentellous, dark grey when dry, midrib prominulous beneath, lateral nerves invisible; young leaves almost glabrous; petiole scarcely 1 mm long; stipules narrowly subulate, acute, spreading, fuscous, persistent, 1·5 mm long. Male flowers unknown. Female flowers solitary, axillary; tepals green, ± herbaceous, erect, narrowly ovate, rigid, minutely pilosulous, acute, 2–3 mm long, canaliculate; ovary globose, finely tomentellous; styles thick, entire, erect, connivent, 1 mm long. Capsule (*teste* Muell. Arg.) globose, ashy-tomentose.

NT (DG)—Endemic. Only known from the type collection from the Gulf of Carpentaria.

Resembles a finely pubescent form of *S. ochrophyllus*, but without the ochraceous colouring, and with more shortly pedicelled flowers with narrow tepals.

Sauropus thesioides (*Hj. Eichl.*) *Airy Shaw* comb. nov.

Glochidion thesioides Hj. Eichler, Suppl. Black's Fl. S. Austr.: 210 (1965), *nom. nov. pro*—

Phyllanthus thesioides Benth., Fl. Austr. 6: 98 (1873); *non* Muell. Arg. (1863). Syntypes: Q. (MO): near Brisbane, *C. Prentice* (♀); N.S.W., Lachlan River, *L. Moreton* (♂) (MEL).

An erect plant with leafy stems to 30 cm long from a woody rootstock, Leaves oblong-oblanceolate, 15–32 × 2–14 mm, long-cuneate to the base. obtuse or subacute at apex, smooth, glabrous, thin, margin narrowly reflexed, green or yellowish when dry, subsessile, stipules subulate, acute, exceedingly minute, almost obsolete. Flowers dioecious, solitary, axillary. Males narrow-infundibular, glabrous, 3–4 mm long; pedicel 2 mm long; tepals short, rounded, 1 mm long; inflexed, the inner smaller than the outer; anthers connate, over half the length of the staminal column, connectives scarcely produced. Females subglobose or shortly obovoid, 1·5 mm long, on longer pedicels, glabrous, pedicel 4–5 mm long; ovary glabrous, styles free, short, thick, entire, just exserted from perianth. Capsule not seen.

Q (BT, DD, WB, MO)—Endemic.

Herb with many slender erect stems 15–30 mm tall from a thick root or short rhizome, locally common on disturbed ground, on dry slopes, or on paths and clearings in eucalypt forest, or on light grey podsol or granite.

Sauropus trachyspermus (*F. Muell.*) *Airy Shaw* comb. nov.

Phyllanthus trachyspermus F. Muell. in Trans. Phil. Soc. Vict. 1: 14 (1855) & in Hook. Kew Journ. & Kew Garden Misc. 8: 210 (1856). Type: N.S.W./V., junction of Darling and Murray Rivers, *F. Mueller* (MEL).

P. rhytidospermus F. Muell. ex Muell. Arg. in Linnaea 34: 70 (1865) & in DC., Prodr. 15(2): 327 (1866); Benth., Fl. Austr. 6: 100 (1873). Type: N.T.: Depôt Creek, Upper Victoria River, *F. Mueller* (MEL).

Glochidion rhytidospermum (Muell. Arg.) Hj. Eichl., Suppl. J. M. Black's Fl. S. Austr. ed. 2: 210 (1965); C. A. Gardner ex Beard, W. Austr. Plants: 58 (?1967).

G. trachyspermum (F. Muell.) Hj. Eichl., l.c. (1965).

Small bushy plant from woody rootstock, with numerous crowded erect or flexuous glabrous leafy stems to 20 cm long, occasionally annual (*Bentham*). Leaves oblong, 5–12 × 2–4 mm, erect or ascending, thin, glabrous, glaucescent, base rounded, apex rounded-mucronate or acute, sometimes minutely asperulous beneath; stipules linear, acute, 2 mm long, spreading. Flowers dioecious. Males not seen. Females very small, shortly pedicelled; tepals oblong, nearly 2 mm long, acute, erect, spreading at the apex, glabrous; ovary glabrous, with 3 very short spreading apically bifid styles. Capsule ovoid, smooth, glabrous, 3–4 mm long, subacute; pedicel 2 mm long; seeds strongly and acutely muricate-asperous.

NT (DG, VD, CA), **Q** (GS, ML, PC, WG), **NSW, SA**—Endemic.

The widespread occurrence of this plant suggests that it is perhaps becoming distributed by artificial means.

The opportunity is taken to transfer also two non-Australian species to *Sauropus*:

Sauropus bacciformis (*L.*) *Airy Shaw* comb. nov.

Phyllanthus bacciformis L., Mant. Pl.: 294 (1767). Type: [S India] Tranquebaria, *Koenig*.

P. racemosus L. f., Suppl.: 415 (1781).

Agyneia impubes Vent., Descr. Pl. Jard. Cels: 23, tab. 23 (1800).

A. bacciformis (L.) Juss., Euphorb. Gen. Tent.: 24, t. 6 (1824); Muell. Arg.: 15(2): 238 (1866); Pax & Hoffm. xv: 213 (1922).

A. phyllanthoides Spreng., Syst. 3: 19 (1826).
Emblica racemosa (L.f.) Spreng., *l.c.*: 20 (1826).
Diplomorpha herbacea Griff., Notul. 4: 479 (1854).
D. bacciformis (L.) Kuntze, Rev. Gen. Pl.: 693 (1891).
Phyllanthus goniocladus Merr. & Chun in Sunyatsenia 2: 260, t. 51 (1935).
Synostemon bacciformis (L.) Webster in Taxon 9: 26 (1960), *in adnot.*; Backer & Bakh. f.: 471 (1963); Airy Shaw in K.B. 26: 343 (1971).

MAURITIUS, INDIA, CEYLON, S CHINA, SE ASIA, W MALESIA (exc. Sumatra, Philippines), CELEBES. Especially on sandy beaches and in other saline situations.

Sauropus sphenophyllus (*Airy Shaw*) *Airy Shaw* comb. nov.

Synostemon sphenophyllus Airy Shaw in K.B. 33: 37 (1978). Type: *Henty & Katik* NGF 38763 (BRI).

NEW GUINEA. Papua: Western District; Morehead subdistr.: shrub of 90 cm in woodland on low hill.

Sebastiania *Spreng.*

A large mostly New World genus (90–100 spp.) of very diverse habit, separable only with difficulty from *Excoecaria*, except by the presence of a caruncle on the seed. The only Australian species is, however, immediately distinguishable from the *Excoecaria* spp. by its subherbaceous habit and shortly spinous capsule.

Sebastiania chamaelea (*L.*) *Muell. Arg.*: 1175 (1866); Benth.: 151 (1873); Bailey: 1456 (1902); Pax & Hoffm. v: 116 (1912); Ewart & Davies: 168 (1917); Gardner: 72 (1931); Backer & Bakh. f.: 498 (1963); Airy Shaw in K.B. 26: 339 (1972) & K.B. Add. Ser. IV: 195 (1975); George & Kenneally in Kabay & Burbidge, Biol. Surv. Drysd. River Nat. Park: 55 (1977). Type: 'Habitat in India', without collector's name (LINN).

Tragia chamaelea L., Sp. Pl.: 981 (1753).
Elachocroton asperococcus F. Muell. in Hook. Journ. Bot. & Kew Garden Misc. 9: 17 (1857). Type: N.T., 'in locis sterilioribus ad flumen Victoriae', *F. Mueller* (MEL).
Stillingia chamaelea (L.) Baill., Ét. Gén. Euphorb.: 516 (1858).
S. asperococca (F. Muell.) Baill., *l.c.*: 517 (1858).
Excoecaria chamaelea (L.) Baill. in Adansonia 6: 323 (1866).

WA (K); **NT** (DG, VR, BT, CA); **Q** (BK, CK, NK)—India and Ceylon to S China, W Malesia (except Philippines), and Solomon Is.

Many-stemmed annual herb or perennial subshrub to 1 m tall, on granite outcrops, on grey stony slopes, on a sandy alluvial rise, in open Eucalypt forest with *Callitris* on sandy soil, behind fore-dunes near beach, in mixed open forest on lateritic outcrops, with *Eucalyptus* and *Acacia* on deep red sandy soil, in flat sandy soil with eucalypt, acacia and spinifex, in messmate savanna-forest on hard lateritic ridges, in brown sandy soil near creek, on sandy foreshore, in forest or deciduous vine thicket, up to 800 m.

The short slender spikes of minute male flowers and the somewhat oblong capsule with two rows of orange teeth on the back of each loculus are unmistakable features of this plant.

The apparent absence of this plant from the Philippines, Celebes, Lesser Sunda Islands and New Guinea is remarkable.

Securinega *Juss.*

Shrubs, closely related to *Phyllanthus*, differing principally in the dioecious flowers and in the presence of a conspicuous pistillode in the males; even closer to *Margaritaria*, but differing in the non-deciduous habit, 5 sepals, 3–5 stamens and presence of a pistillode.

1. Lateral branches short, often spinescent; leaves small, mostly under 2 cm long, thin, spathulate-obovate, often truncate-emarginate **S. leucopyrus**
1. Lateral branches longer, very rarely spinescent; leaves larger, somewhat thicker, up to 9·5 × 6 cm, mostly elliptic or obovate-oblong, mostly subacute, occasionally emarginate:
   2. Nervation of leaves prominent and reticulate beneath; average size of leaves larger . . . . **S. melanthesoides**
   2. Nervation smooth and inconspicuous beneath; average size of leaves smaller (doubtfully native) . . . . **S. virosa**

Securinega leucopyrus (*Willd.*) *Muell. Arg.*: 451 (1866); Benth.: 116 (1873); Airy Shaw in K.B. 25: 493 (1971) & K.B. Add. Ser. IV: 340 (1975). Type: 'Habitat in India orientali', *Klein* (B†).

Flueggea leucopyrus Willd., Sp. Pl. 4: 757 (1805); Bailey: 1426 (1902); Ewart & Davies: 162 (1917).

Securinega virosa var. *australiana* Baill. in Adansonia 6: 334 (1866). Syntypes: Q., Gilbert River, *F. Mueller* (P, MEL); Bowen River, *Bowman* 246 (P, MEL); *sine loc.*, 1863, *Dallachy* 117 (P, MEL); Rockhampton, *Thozet* 29 (P, MEL).

Q (CK, NK, SK, PC)—India, Ceylon, SE Asia.

Shrub or small tree of 3–5 m, only habitat noted: amongst boulders on limestone outcrop up to 360 m.

Leaves generally spathulate-obovate; lateral branchlets often spinescent. The taxonomic status of this plant is uncertain. The fact that it seems to appear sporadically throughout the range of *S. virosa* and *S. melanthesoides* could suggest the possibility that it represents an ecotypic variation of these species occasionally called forth by peculiar local circumstances. Cultivation experiments might be interesting.

Securinega melanthesoides (*F. Muell.*) *Airy Shaw* in K.B. 31: 352 (1976).

Leptonema melanthesioides [sic] F. Muell. in Hook. Journ. Bot. & Kew Garden Misc. 9: 17 (1857). Type: N.T., Victoria River, 1856, *F. Mueller* (MEL, K).

Fluggea melanthesoides [sic] (F. Muell.) F. Muell. in Trans. Bot. Soc. Edinb. 7: 490 (1863).

var. **melanthesoides**

[*Securinega obovata* sec. Muell. Arg. 449 (1866); Baill. in Adansonia 6: 333 (1866); Benth.: 115 (1873); Specht: 462, 492, 502 (1958); *vix* (Willd.) Muell. Arg.]

[*Flueggea microcarpa* sec. Bailey: 1426 (1902); Ewart & Davies: 162 (1917); Gardner: 72 (1931); *vix* Bl.]

F. novaguineensis Valet. in Bull. Dép. Agric. Indes Neerl. 10: 26 (1907), *nomen*; Hallier in Meded. Rijks Herb. Leiden 36: 5 (1918), *in obs.* Type: W New Guinea, Merauke, 1907, *Versteeg* 1884 (BO, K).

[*F. virosa* sec. J. J. Sm. in Nova Guinea 8: 221 (1910); Domin: 878 [324] (1927) (incl. f. *reticulata* Domin, *l.c.*, *in obs.*); *vix* (Roxb.) Baill.]

WA (K); **NT** (VR, DG, CA); **Q** (BK, CK, NK)—New Guinea.

Diffuse, straggly shrub to 3 m high and 5 m wide, in open eucalypt forest, open mixed shrubby community on limestone outcrop, or stony surface of volcanics, base of sandstone hill in silty sand, on dark clayey or sandy loam, escarpment gully rain-forest on sandstone, in paper-bark forest on river bank, in savanna forest adjacent to rain-forest, or in sandy soil in littoral forest, up to 450 m.

The strongly elevate reticulate venation and generally larger leaves distinguish this taxon from *S. virosa*.

var. **aridicola** (*Domin*) *Airy Shaw* in Muelleria 4: 213 (1980).

Flueggea virosa var. *aridicola* Domin: 878 [324] (1927). Type: Q. (BK), dry hills near Cloncurry, 1910, *Domin* (PR).

Q (BK)—Endemic.

A xerophilous form, with stout branches and sometimes persistent subspinescent branchlets, and small, glaucous, coriaceous leaves up to 2·7 cm long.

Securinega virosa (*Roxb. ex Willd.*) *Baill.* in Adansonia 6: 334 (1866) (quoad synon. tantum, excl. descr.); Pax & Hoffm. in Engl. & Harms, Pflanzenf. ed. 2, 19c: 60 (1931); Backer & Bakh. f.: 466 (1963); Airy Shaw in K.B. Add. Ser. IV: 196 (1975), *q.v.* for detailed synon., & in K.B. 31: 356 (1976), *in obs.* Type: 'In India orientali', Circars, *Klein* (B†); Calcutta Botanic Garden, *Roxburgh* (B†, K).

Phyllanthus virosus Roxb. ex Willd., Sp. Pl. 4: 578 (1805).

NT (DG)—Tropical Africa and Asia eastward to China and Japan, and with varying frequency through Malesia to Celebes, the Moluccas and the Lesser Sunda Is.

Small spreading shrubby tree of 4–5 m; at very low altitude.

Differs from *S. melanthesoides* in the much less conspicuous and less elevate-reticulate venation of the generally smaller leaves.

The native status of *S. virosa* in Australia is very doubtful and requires confirmation. The plant is so far known only from the Darwin Botanic Garden and from a roadside near the sea a few miles distant.

Suregada *Roxb. ex Rottl.*

Shrubs or small trees, glabrous throughout, the branchlets remaining green when dry. Leaves alternate, elliptic, chartaceous to coriaceous, entire (or sometimes toothed—not in Australia), penninerved, sometimes strongly reticulately nerved, with a conspicuous pellucid dot in each areole, shortly petioled; stipules very small, connate, early caducous, often leaving a conspicuous scar. Flowers dioecious; inflorescences leaf-opposed, fascicled or very shortly cymose, often gummy when young. Male flower: sepals 5, suborbicular, strongly imbricate; stamens numerous, exserted, the receptacle bearing numerous small glands between the filaments; pistillode 0. Female flower: sepals as in the male; disk annular; ovary ovoid, 2–3-locular, glabrous; styles 2–3, very short, bifid, spreading. Fruit capsular or subdrupaceous, globose or shallowly 2–3-lobed, mostly smooth, tardily dehiscent; seeds subglobose, testa ± fleshy in the living state.

Suregada glomerulata (*Bl.*) *Baill.*, Ét. Gén. Euphorb.: 396 (1858); Airy Shaw in K.B. Add. Ser. IV: 198 (1975) & in K.B. 31: 397 (1976). Type: Java, 'in sylvis humilioribus insularum Javae ac Nusae Kambangae', *Blume* (BO).

Erythrocarpus glomerulatus Bl., Bijdr.: 605 (1825).
Gelonium glomerulatum (Bl.) Hassk., Cat. Hort. Bogor.: 237 (1844); Muell. Arg.: 1128 (1866); J. J. Sm.: 594 (1910); Pax & Hoffm. iv: 18 (1912); Backer & Bakh. f.: 497 (1963).
G. papuanum Pax in Engl. iv: 20 (1912); Airy Shaw in K.B. 16: 367 (1963). Type: W New Guinea, *Versteeg* 1553 (holotype B†; isotype K).
Doryalis macrodendron Gilg in Engl., Bot. Jahrb. 55: 283, fig. 5 (1918); cf. Airy Shaw in K.B. 16: 367 (1963). Type: NE New Guinea, Pfingstberg, *Ledermann* 7400 (B†).

NT (DG)—Throughout Malesia to New Guinea.

Small tree to 5 m tall, in rain forest and marginal scrub and by a permanent creek at 75 m. Fig. 6B.

Quite glabrous. Branchlets marked with the prominent scars of the caducous connate stipules. Leaves elliptic or narrowly cuneate-obovate, up to 18 × 6 cm, decurrent at the base into the very short petiole, obtuse or rounded at the apex, smooth and rather shining, green when dry, with translucent areoles between the ultimate nerves. Inflorescences short, fascicled, conspicuously leaf-opposed, often very gummy. Capsule globose, orange-red, 10–12 mm diam., pedicel 5 mm long.

So far known only from two localities in Arnhem Land.

SYNOSTEMON *F. Muell.* See **Sauropus** *Bl.*

Tragia *L.*

Mostly twining or scrambling shrubs, frequently stinging. Leaves alternate, very varied in outline, often ± ovate; stipules triangular or lanceolate. Inflorescences racemose, terminal or leaf-opposed, rarely axillary, the flowers mostly ♂, with 1–few basal ♀. Male flower: sepals 3(–5), valvate; disk-glands few or 0; stamens mostly 3, alternisepalous, rarely up to 50, filaments very short, anthers extrorse or introrse; pistillode very small, often trifid. Female flowers: sepals (3–)6, entire or pinnatifid, sometimes accrescent; disk 0; ovary 3-locular, loculi 1-ovulate; styles simple, connate below into a column. Capsule tricoccous; seeds globose, ecarunculate, sometimes puberulous.

Tragia novae-hollandiae *Muell. Arg.* in Linnaea 34: 180 (1865) & in DC.: 929 (1866); Baillon in Adansonia 6: 320 (1866); Benth.: 138 (1873); Moore & Betche: 78 (1893); Bailey: 1453 (1902); Pax & Hoffm. ix–xi: 44 (1919); Airy Shaw in K.B. 23: 117, *in obs.* (1969). Syntypes: N.S.W., Clarence River, *F. Mueller*; Q., Keppel Bay, *F. Mueller* (G–DC).

Q (NK, PC, MO); **NSW** (NC)—Endemic.

Slender twining herb or scrambler over trees in moist spots or in bush at the base of cliffs at unknown altitude.

Clothed with a sparse pubescence of very short decurved crispulous hairs and longer straight setulose urticating hairs. Leaves triangular or ovate-triangular, cordate with a broad sinus at the base, acute at the apex, coarsely and acutely triangular-serrate or rarely subentire, 2–9 × 1–6 cm, herbaceous-membranaceous, pale beneath; petiole up to 3·5 cm long, very slender; stipules deltoid-subulate, 2–3 mm long, brown when dry. Inflorescences short, androgynous, mostly leaf-opposed, up to 6 cm long in fruit. Male flower with 5 valvate calyx-segments, ± reflexed; stamens 3–5, with subsessile erect extrorse anthers; pistillode broad, flat. Female flower with 6 imbricate calyx-segments; ovary shortly pilose, styles connate more than half-way, simple, recurved at the apex. Capsule deeply tricoccous, 8–11 × 4–5 mm, setulose or subglabrous.

This species, the only Australian representative of the tribe *Plukenetiëae*, in some respects forms a bridge between the more typical species of *Tragia* and the monotypic genus *Pachystylidium* Pax & Hoffm., of India, Southeast Asia and western Malesia; *cf.* Airy Shaw, *l.c. supra* (1969).

Trigonostemon *Bl.*

Trees or shrubs, without stellate hairs. Leaves alternate or subopposite, entire or denticulate, penninerved, often triplinerved at the base; stipules mostly minute and subulate or obsolete. Inflorescences axillary or terminal, occasionally cauliflorous, monoecious, ♂ or ♀ or ♂♀, variously cymose, thyrsiform or apparently racemose, often with conspicuous bracts. Male flower: sepals 5, very shortly connate, imbricate; petals 5, free, mostly exceeding the sepals, occasionally bilobed, often brightly coloured (yellow, orange, red, black-crimson, etc.); disk-glands 5, often united into a ring or cup; stamens 3 or 5 (very rarely 13); filaments connate or free and spreading above, anthers extrorse, the connectives sometimes much produced and horn-like; pistillode 0. Female flower: sepals much as male, occasionally fringed with capitate glands, rarely accrescent in fruit; petals as male, but sometimes differently coloured, occasionally 0; disk-glands free or variously united; ovary 3-locular, styles simple or once or twice bifid, spreading. Capsule tricoccous, smooth or verruculose; seeds trigonous-ovoid.

Trigonostemon inopinatus *Airy Shaw* in K.B. 31: 396 (1976). Type: Q., Cawley State Forest, near Cathu, 1965, *Webb & Tracey* 7762 (BRI).

Q (SK)—Endemic.

Small understorey tree of 3–6 m, in mixed notophyll vine forest on soils derived from granite, at 600–800 m.

Young parts densely shortly fulvous-villous, soon glabrescent. Leaves elliptic or obovate-elliptic, 4–9 × 2–4 cm, mostly obtuse at the apex, margin entire or sinuate-crenate towards the apex, sparsely adpressed-pilose on both surfaces; petiole up to 2·5 cm long; stipules obsolete. Male inflorescence shortly cymose, shortly adpressed-pilose, peduncle 1 cm long, bracts 3–6 mm long, pedicels 5–6 mm long. Sepals 5–6, narrowly subulate, 3–5 × 1 mm. Petals 5, contorted, broadly obovate, 4 mm long, 2·5–3 mm diam., cream with a red base or entirely blackish red. Stamens 3, on a thick staminal column. Female inflorescence unknown, apparently 1-flowered. Capsule subglobose, 13 mm diam., smooth, subglabrous, borne on a pedicel 2·5 cm long and subtended by oblong foliaceous sepals 6–11 × 1·5–4 mm; styles very shortly flabellate, 1 mm long.

Apparently not uncommon in Cawley State Forest, between Mackay and Proserpine, but not yet known elsewhere.

Whyanbeelia *Airy Shaw & Hyland*

Distantly related to *Dissiliaria*, from which it differs in its monoecious flowers, often with males and females in the same inflorescence, in the laxly cymose inflorescences, in the numerous (± 50) stamens, in the male disk consisting of 3 short radiating rows of glands concealed among the stamens, in the acute narrowly subulate female sepals, and in the short subulate glands of the female disk.

Whyanbeelia terrae-reginae *Airy Shaw & Hyland* in K.B. 31: 376 (1976). Type: Q., Timber Reserve 55 Whyanbeel, *Hyland* 3052 RFK (K).

Q (CK)—Endemic.

Tree to 20 m tall, in rain-forest at 150–200 m. Fig. 6D.

Dioecious, the young parts shortly pubescent. Leaves opposite, narrowly ovate, 6–16 × 2·5–6 cm, rounded at the base, gradually long-attenuate to the apex, entire, somewhat shining, glabrous or almost so; petiole 4–9 mm long; stipules apparently obsolete. Inflorescences arising from distal axils, loosely cymose, up to 6 cm long, entirely male or with a few female flowers intermixed, minutely puberulous. Male flowers borne on a pedicel up to 9 mm long; sepals 3 + 3, the outer ovate, 1–1·5 × 1 mm, the inner suborbicular, 2 mm diam., all dorsally pilosulous, tightly adpressed to the stamens; stamens 50–55, inserted on a densely pilosulous hemispherical receptacle, in the bud stage forming a dense broadly 3-lobed mass; filaments 2 mm long at anthesis, anthers small, subglabrous; disk formed of 3 radiating rows of glands amongst the stamens, the glands angular and fleshy; pistillode none. Female flowers borne on a 9 mm long pedicel; sepals 6, narrow-subulate, ± equal, acute, disk-segments subulate, 1–2 mm long, glabrous; ovary ovoid, 3 × 2 mm, densely pubescent; styles 3, simple, recurved, 3–4 mm long. Capsule subglobose, up to 2·2 cm diam., puberulous.

So far known only from a very few trees in the Daintree River valley.

Appendix

STILAGINACEAE

Antidesma *L.*

Shrubs or small trees, dioecious. Leaves alternate, entire, penninerved, the main nerves forming conspicuous arcuate anastomoses, mostly short-petioled. Inflorescences simply or compoundly racemose or spicate, sometimes cauliflorous; flowers very small, solitary in the axils of mostly small bracts. Male flower: calyx cupular, 3–5-toothed or deeply 3–5-lobed, the lobes imbricate; petals 0; disk pulviniform or extrastaminal, composed of free or ± connate glands; stamens 2–5 (rarely 10), arising within the disk or between the disk-glands or in excavations of the disk, usually much exserted, the anther loculi distinct, borne on a thickened connective; pistillode usually small but distinct. Female flower: calyx ± as male; hypogynous disk annular; ovary 1- or occasionally 2-locular, loculi 2-ovulate, styles 3(–5), short, bilobed, terminal or lateral. Fruit drupaceous, small or very small, oblique or symmetrical, often compressed, indehiscent, red or black when ripe, endocarp laxly reticulate-foveolate.

1. Leaves less than 2·5 cm long, obovate; inflorescences short and slender; drupes small, elliptic, flattened, glabrous; styles subterminal **A. parvifolium**
1. Leaves over 2·5 cm long; inflorescences less slender:
  2. Leaves often deeply and irregularly sinuate; drupes compressed, 2·5–3·5 mm long . . . . **A. sinuatum**
  2. Leaves entire, not sinuate:
    3. Leaves broadly rounded or emarginate at apex, or occasionally shortly acute; drupe mostly small, elliptic, with terminal styles, sometimes larger, sometimes bilocular and broader than long **A. ghaesembilla**
    3. Leaves not broadly rounded or emarginate at apex, glossy on the upper surface:
      4. Leaves broad, up to 20 × 10 cm; drupes larger . . . **A. bunius**
      4. Leaves narrow, up to 10 × 4 cm; drupes smaller . . . **A. erostre**

Antidesma bunius (*L.*) *Spreng.*, Syst. Veg. 1: 826 (1825); Muell. Arg.: 262 (1866); Benth.: 86 (1873); Bailey: 1433 (1902); J. J. Sm.: 270 (1910); Pax & Hoffm. xv: 160 (1922); Backer & Bakh. f.: 458, 460 (1963); Airy Shaw in K.B. 26: 353 (1971) & in K.B. Add. Ser. IV: 209 (1975). Type: 'Habitat in India'; '*Bunius sativus* Rumph. amb. 3. p. 204, t. 131'.

Stilago bunius L., Mant.: 122 (1767).
A. dallachyanum Baill. in Adansonia 6: 337 (1866); Benth.: 85 (1873) (excl. specim. 'Port Essington, *Armstrong*'); Bailey: 1432 (1902); Pax & Hoffm. xv: 120 (1922); Hyland: 62 (1971); **synon. nov.** Type: Q., Rockhampton, 'Dalrymph Cape', 1864, *Dallachy* (P, MEL).

Q (CK, NK)—S India, Ceylon, E Himalaya, SE Asia, S China and throughout Malesia (excl. Malaya and mainland Borneo) to New Guinea.

Shrub or small tree in rain-forest or gallery rain-forest at 50–450 m.

Shrub or tree to 10 m high, with smooth, glossy, glabrous, somewhat laurel-like leaves 7–15 cm long. Inflorescences usually robust, elongate, almost simple, rufous-pubescent. Male flowers sessile, cupuliform, with 4 stamens, filaments elongate and relatively thick, anthers large for the genus. Infructescence very robust, drupes large, compressed-ovoid, pedicellate, with terminal styles.

Bentham's records of *A. bunius* and *A. dallachyanum* were based on a single collection each, both of Dallachy from Rockingham Bay. He also referred to 'two other species or varieties of *Antidesma* with the large glabrous leaves of *A. bunius*,' one with large black fruits and the other with large white fruits. I have very little doubt that all these plants represent variations of *A. bunius*.

Baillon cites the type of *A. dallachyanum* as '*Dallachy* (1864), Rockhampton, Dalrymph Cape (herb. F. Muell.!)'. The apparent isotype at Kew, however, is annotated in Bentham's handwriting as 'Rockingham Bay, Dallachy', and is so cited by him in the Flora. This discrepancy will need to be resolved in due course. The two localities, Rockhampton and Rockingham Bay, are very liable to be confused; cf. 'Rockhampton Bay' under *Cleistanthus apodus* in Benth., Fl. Austr. 6: 122 (1873). I have been unable to trace any locality like 'Dalrymph Cape' in the region of Rockhampton or elsewhere, but of course Mt Dalrymple and Dalrymple Heights are well-known localities to the west of Mackay.

The Kew sheet appears to contain a mixture. The lowermost piece, mounted horizontally, with smaller leaves, dull below and folded longitudinally, is almost certainly a member of the variable *A. ghaesembilla* complex. But I believe Bentham was mistaken in asserting (*l.c.*) that *A. dallachyanum* is 'closely allied to *A. ghaesembilla*'. In Pax & Hoffmann's monograph *A. bunius* and *A. ghaesembilla* are in fact—I believe rightly—widely separated; in foliage, inflorescence and fruit there are profound differences.

Antidesma erostre *F. Muell. ex Benth.*: 87 (1873); Bailey, Syn. Queensl. Fl.: 472 (1883) & Queensl. Fl. 5: 1433 (1902); Pax & Hoffm. xv: 166 (1922); Hyland: 43 (1971); Airy Shaw in K.B. 28: 279 (1973), *q.v.* Type: Q., Rockingham Bay, *Dallachy* (MEL).

Q (CK, NK, SK)—Papua.

Shrub or tree to 20 m high, locally common in rain-forest on red basaltic soils up to 1050 m.

The glossy, elliptic or oblong-elliptic, firmly chartaceous leaves, up to 12 × 4 cm, narrowed almost equally to base and apex, with distinctly prominulous venation, especially beneath, are diagnostic for this species. Young shoots, inflorescence-rhachis and ♀ calyx (at least at base) minutely puberulous; plant otherwise glabrous, including ♂ calyx and ovary. Male flowers sessile; calyx broadly cupular, shortly 6-lobed; disk extra-staminal, 4–6-lobed; pistillode very short, glabrous. Female flowers shortly pedicelled; calyx shortly cylindric, shortly 5–6-lobed; disk deeply annular. Drupe small, ellipsoid, flattened, with terminal styles.

The species may well be related to the Philippine *A. digitaliforme* Tul., as I suggested in 1973 (*l.c. supra*), but certainly not to *A. schultzii* Benth., which it is now clear is part of the *A. ghaesembilla* complex (*vide infra*). There may also be some connexion with *A. sinuatum* Benth.

Antidesma ghaesembilla *Gaertn.*, Fruct. 1: 189, t. 39 (1788); Muell. Arg.: 251 (1866); Baill. in Adansonia 6: 337 (1866); Benth.: 85 (1873); Bailey: 1432 (1902); J. J. Sm.: 287 (1910); Pax & Hoffm. xv: 155 (1922); Gardner: 72 (1931); Backer & Bakh. f.: 458 (1963); Airy Shaw in K.B. 26: 353 (1972) & K.B. Add. Ser. IV: 211 (1975); George & Kenneally in Kabay & Burbidge, Biol. Surv. Drysd. River Nat. Park: 54 (1977). Type: 'e collect. sem. hort. lugdb.' (L).

Antidesma schultzii Benth.: 86 (1873); Ewart & Davies: 166 (1917); Pax & Hoffm. xv: 134 (1922); **synon. nov.** Syntypes: N.T., Port Darwin, *Schultz* 610 & 743 (K).

WA (K); **NT** (DG, VR); **Q** (CK)—W Himalaya and Ceylon to S China and SE Asia, and throughout Malesia to New Guinea and the Bismarck Archipelago.

Shrub of 3–5 m, in open forest on lateritic sandy soil, or at edge of dry creek in eucalypt woodland, or in stabilized dunes, at 75 m.

Leaves mostly broadly elliptic, 2·5–7·5 × 1·5–5(–10) cm, rounded or emarginate at the apex, broadly cuneate to rounded or even cordate at the base. Inflorescences (both male and female) usually much-branched and densely ochraceous-pubescent. Drupe small, ellipsoid, with terminal styles.

The variation of *A. ghaesembilla* in Australia needs further study from abundant material. On the basis of two collections from Darwin, *Schultz* 610 and 743 (both in female flower, in which the ovary was quite glabrous), Bentham (*l.c.*) segregated a supposed new species, *A. schultzii* Benth.; but two further collections from the same locality, *Schultz* 694 and 748 (both in male flower), he retained in *A. ghaesembilla*.

I have so far failed to detect any convincing correlation of the ovary character with any other of the many variables evident in the Australian material, such as indumentum in general, leaf-size and shape, petiole-length, branching of inflorescence, sepal-shape, fruit-size, etc. At present it does not seem possible to recognize more than one taxon in this variable complex.

Antidesma parvifolium *F. Muell.*, Fragm. 4: 86 (1864); Benth.: 86 (1873); Bailey: 1433 (1902); Ewart & Davies: 166 (1917). Syntypes: Q., Port Denison, *Fitzalan*, *Dallachy* (MEL).

NT (DG); **Q** (CK, NK, PC)—Endemic.

Shrub of 2–3 m, in broadleaf scrub associated with small sandstone outcrop, or in sandstone gorge, or in stony soil on hilltop, or in open forest or in deciduous vine thicket, up to 880 m.

The very small obovate leaves, always less than 2·5 cm in length, are sufficient to distinguish *A. parvifolium* from the remaining species; the leaves are cuneate at the base and rounded or emarginate at the apex. Both male and female inflorescences short and slender; male flowers sessile, with suborbicular sepals and 3–4 stamens, glabrous except for the puberulous pistillode; females shortly pedicelled, with ovate, mostly acute, glabrous, erose-margined sepals; drupes small, elliptic, flattened, glabrous, with terminal or subterminal styles.

The only close relative of *A. parvifolium* seems to be *A. spatulifolium* Airy Shaw (K.B. 23: 283 (1969)), of southern Papua.

Antidesma sinuatum *Benth.*: 87 (1873); Bailey: 1433 (1902); Pax & Hoffm.: xv: 167 (1922). Type: Q., Rockingham Bay, *Dallachy* (K).

Q (NK)—Endemic.

No ecological information available.

The curious sinuate-margined leaves of this plant are very distinctive, but they give the impression of something abnormal, as though the plant were suffering from some constitutional disturbance. I have even wondered whether Dallachy's unique fruiting specimen might represent a hybrid between *A. bunius* (*A. dallachyanum*) and *A. erostre*, since the foliage (apart from the sinuate margin) is somewhat intermediate between that of the two species, and all three taxa probably originated from the same region, Rockingham Bay.* The fruits of *A. sinuatum* seem, however, to be normally developed, giving no outward evidence of hybrid sterility. The matter can probably only be solved by the collection of more abundant material.

(Index overleaf)

* See under *A. bunius*, p. 693 *supra*. Bentham's version of the name of the type locality of *A. dallachyanum* is more likely to be correct than Baillon's, which in any case contains at least one obvious error ('Dalrymph' for Dalrymple).

INDEX

Index to genera and species (Introduction and keys excepted). Accepted names in **bold,** synonyms in *italics*; names merely mentioned in discussion in roman.